Galois Fields and Galois Rings Made Easy

To my granddaughter Éloïse Kibler Blau

Galois Fields and Galois Rings Made Easy

Maurice R. Kibler

First published 2017 in Great Britain and the United States by ISTE Press Ltd and Elsevier Ltd

ISTE Press Ltd
27-37 St George's Road
London SW19 4EU
UK

www.iste.co.uk

Elsevier Ltd
The Boulevard, Langford Lane
Kidlington, Oxford, OX5 1GB
UK

www.elsevier.com

For information on all our publications visit our website at http://store.elsevier.com/

British Library Cataloguing-in-Publication Data
A CIP record for this book is available from the British Library
Library of Congress Cataloging in Publication Data
A catalog record for this book is available from the Library of Congress
ISBN 978-1-78548-235-9

Printed and bound in the UK and US

Contents

Acknowledgments

I am indebted to Natig Atakishiyev, Mohammed Daoud, Serge Perrine, Michel Planat, Metod Saniga and Bernardo Wolf, as well as, last (but not the least), my student Olivier Albouy for numerous discussions, to Apostol Vourdas for discussions and e-mail correspondence on Galois quantum mechanics, to Bruce Berndt, Ron Evans and Philippe Langevin for e-mail correspondence on quadratic Gauss sums, to Markus Grassl, Arthur Pittenger and Stefan Weigert for e-mail correspondence on MUBs, and to Philippe Caldero for a reading of the manuscript and providing valuable comments.

Finally, I am grateful to my wife Gloria for her patience and continual encouragement in the course of writing this book.

Preface

This book constitutes an elementary introduction to rings and fields, especially Galois rings and Galois fields, with regard to their application to the theory of quantum information.

Since the 1930s, the theory of groups has been widely used in many domains of physical sciences (elementary particle and nuclear physics, atomic and molecular physics, condensed matter physics, theoretical and quantum chemistry). In contrast, the theory of rings and fields, which comes immediately after group theory in the hierarchy of abstract algebra, is less well known to physicists and chemists. Of course, fields with an infinite number of elements like the field of real numbers, the field of complex numbers and, to some extent, the field of quaternions are all well known in the physical sciences. Similarly, infinite rings, such as the ring of integers and the ring of square matrices, are of common usage in physics and chemistry. However, finite rings and finite fields (largely used in pure mathematics and in the classical theory of information) are relatively unknown to physicists and chemists - despite their potential utility for the quantum theory of information, having been recognized in the 1990s.

The existing literature on rings and fields is primarily mathematical. There are a great deal of excellent books on the theory of rings and fields written by and for mathematicians, but these can be difficult for physicists and chemists to access. The present book offers an introduction to rings and fields for students and researchers in physics and chemistry, with an emphasis on their application to the construction of mutually unbiased bases of pivotal importance in quantum information. This book is intended for graduate and undergraduate students and researchers in physics, mathematical physics and quantum chemistry (especially in the domains of advanced quantum mechanics, quantum optics, quantum information theory, classical and quantum computing, and computer engineering). Although the book is not written for mathematicians, given the large number of examples discussed, it may be of interest to undergraduate students in mathematics.

The book is organized as follows. Chapter 1 is devoted to a general discussion of the algebraic structures of rings and fields. Chapters 2 and 3 deal with Galois fields (i.e. finite fields) and Galois rings (i.e. special finite rings) respectively. Chapter 4 is concerned with the construction of mutually unbiased bases in Hilbert spaces of finite dimension; for Hilbert spaces of dimension p^m, with p a prime number and m a positive integer, Galois rings are used for p even and Galois fields for p odd. Finally, for the reader unfamiliar with number theory and group theory, an appendix (Chapter 5) lists some basic results that are necessary for the understanding of the first four chapters. Finally, a list of references (Bibliography) closes the book; this list includes some relevant web links. In sum, the book is divided into two parts: a mathematical part (Chapters 1, 2, 3 and 5) and a physical part (Chapter 4). Some further specification of the two parts is in order.

The presentation and the pedagogy of the mathematical part of this book differ in many respects from those encountered in books dealing with pure mathematics. This part may be considered as a collection of definitions and results (propositions and properties) useful to the practitioner. The emphasis is placed on examples (with numerous repetitions), that sometimes come before the results, rather than on proofs of the results. Nevertheless, some proofs and some basic elements of proof are given. More involved proofs can be found in the books listed in the sections *Mathematical literature: rings and fields* and *Mathematical literature: Theory of numbers* of the bibliography at the end of the book. In this respect, the seminal textbooks by B.C. Berndt, R.J. Evans and K.S. Williams, M. Demazure, L.K. Hua, R. Lidl and H. Niederreiter, and Z.-X. Wan were of invaluable help in writing the present book. The reader is encouraged to consult these books not only for their mathematical content but also for their impressive lists of references and historical notes.

For the physical part of this book, a basic knowledge of quantum mechanics is required. The concept of a set of mutually unbiased bases (MUBs for short) has been the subject of great activity since the end of the

1990s (a nonexhaustive list of references is given in the section *Theoretical physics literature: MUBs*). The concept of MUBs is introduced in Chapter 5 via the theory of quantum angular momentum, i.e. in mathematical terms, via the theory of the Lie algebra of the group SU(2). A complete set of MUBs is constructed in the Hilbert space $\mathbb{C}^p$ for p prime; the case of $\mathbb{C}^{p^m}$, with m positive integer, can be deduced from tensorial product. The interest of Galois fields and Galois rings for the construction of a complete set of MUBs in $\mathbb{C}^{p^m}$ is shown as an application of the principles discussed in Chapters 2 and 3.

To conclude, let us mention that the calculations arising in the numerous examples can be achieved very easily by using a symbolic and numeric programming language such as Maple.

Maurice KIBLER
June 2017

List of Mathematical Symbols

Sets

→ the symbol $\forall$ means *for any*

→ the symbol $\exists$ means *there exists*

→ depending on the context, the vertical bar $|$ may mean *such that* (for example, $\alpha \mid f(\alpha) = 0$ means α is such that $f(\alpha) = 0$) or *divides* (like in $b \mid a$ which means b divides a) or may appear in $\langle x|y \rangle$ (which denotes an inner product in a pre-Hilbert space)

→ $B \subset A$ (or equivalently $A \supset B$) means that the set B is a subset of A; other meanings of $B \subset A$ can be: B is a sub-group of A or B is a sub-ring of A or B is a sub-field of A

→ $A \cup B$ denotes the union of the sets A and B

→ if $B \subset A$, then $A \setminus B$ denotes the set A from which the elements of the set B are missing

→ $\text{Card}(S)$ denotes the cardinal of the set S

Numbers

→ $\mathbb{N}_0 = \{0, 1, 2, \cdots\}$ is the set of non-negative integers, 0 included ($\mathbb{N}_0$ is denoted as $\mathbb{N}$ in the French literature)

→ $\mathbb{N}_1 = \mathbb{N}_0 \setminus \{0\} = \{1, 2, 3, \cdots\}$ is the set of positive integers, 0 excluded ($\mathbb{N}_1$ is denoted as $\mathbb{N}^*$ in the French literature)

→ $\mathbb{Z}$ is the set of integers (including 0): $\mathbb{Z}$ comprises the negative integers, 0 and the positive integers

→ $d\mathbb{Z}$ is the set of integers divisible by $d \in \mathbb{N}_1$

→ $\mathbb{Q}$ is the set of rational numbers

→ $\mathbb{R}$ is the set of real numbers

→ $\mathbb{C}$ is the set of complex numbers

→ $\mathbb{C}^* = \mathbb{C} \setminus \{0\}$ is the set of complex numbers without 0

→ $\mathbb{H}$ is the set of quaternions ($\mathbb{H} \supset \mathbb{C} \supset \mathbb{R} \supset \mathbb{Q} \supset \mathbb{Z} \supset \mathbb{N}_0 \supset \mathbb{N}_1$)

→ $n!$ denotes the factorial of $n \in \mathbb{N}_0$

→ e stands for the basis of Napierian logarithms

→ $\mathrm{i} = \sqrt{-1}$ stands for the pure imaginary number

→ $\overline{z}$ denotes the complex conjugate of $z \in \mathbb{C}$

→ $|z|$ denotes the modulus of $z \in \mathbb{C}$

→ $\mathrm{i}, \mathrm{j}, \mathrm{k}$ are the basic quaternions

→ p is an even or odd prime (the letter p is used for denoting prime numbers)

→ p^m denotes a power of a prime ($m \in \mathbb{N}_1$)

→ $a \mid b$ means that the integer a divides the integer b

→ $a \not| b$ means that the integer a does not divide the integer b

→ $a \equiv b \bmod n$ means $n \mid (a - b) \Leftrightarrow a = b + kn$, where $k \in \mathbb{Z}$; we say that a is congruent to b modulo n (in many places where the context *modulo* is clear, $a \equiv b \bmod n$ is simply noted as $a = b$)

→ the residue of $a \in \mathbb{Z}$ modulo $n \in \mathbb{Z}$ is $b \in \mathbb{Z}$ such that $a = nk + b$ with $n \in \mathbb{Z}$ and $0 \leq b < n$ ($\Rightarrow\ a \equiv b \bmod n$)

→ $\gcd(a, b)$ stands for the greatest common divisor of the integers a and b

→ $(a, b) = 1$ means that the integers a and b are co-prime (their greatest common divisor is 1)

→ δ is the Kronecker symbol (for $m, n \in \mathbb{Z}$, $\delta(n, m) = 0$ if $n \neq m$ and $\delta(n, m) = 1$ if $n = m$)

→ φ denotes the Euler function

→ μ denotes the Möbius function

→ $(\frac{a}{p})$ is the Legendre symbol (a integer and p prime)

→ C_p^k is the Newton binomial coefficient

→ $G(d)$ denotes the usual Gauss sum

Matrices

→ the sign $\otimes$ indicates the tensor product of vectors (in the framework of vector spaces)

→ $\det(A)$ denotes the determinant of the matrix A

→ $\mathrm{tr}(A)$ denotes the trace of the matrix A (tr is reserved for matrices and Tr for rings and fields)

→ $A^\dagger$ stands for the Hermitian conjugate (i.e. transpose + complex conjugate) of the matrix A

→ $[X, Y]$ or $[X, Y]_-$ denotes the commutator of the matrices (or operators) X and Y

→ $[X, Y]_+$ denotes the anticommutator of the matrices (or operators) X and Y

Groups

→ $|G|$ denotes the cardinal (or order) of the group G

→ ker(f) denotes the kernel of the group homomorphism f

→ G/H stands for the quotient group G by its normal sub-group H

→ $G \simeq G'$ means that the groups G and G' are isomorphic (the symbol $\simeq$ is reserved to groups)

→ $G \times G'$ is the direct product of the groups G and G'

→ V is the Klein four-group

→ C_d denotes the cyclic group of order d

→ $\langle a \rangle$ stands for the cyclic group generated by the element a

→ S_n is the symmetric group on n objects

→ A_n is the alternating group on n objects

→ GL$(n, \mathbb{C})$ is the general linear group, in n dimensions, on the field $\mathbb{C}$

→ SL$(n, \mathbb{C})$ is the special linear group, in n dimensions, on the field $\mathbb{C}$

→ O(n, R) is the orthogonal group, in n dimensions, on the field $\mathbb{R}$

→ SO(n, R) is the special orthogonal group, in n dimensions, on the field $\mathbb{R}$

→ U$(n, \mathbb{C})$ is the unitary group, in n dimensions, on the field $\mathbb{C}$

→ SU$(n, \mathbb{C})$ is the special unitary group, in n dimensions, on the field $\mathbb{C}$

Rings

→ the signs $+$ and $\times$ (sometimes $\oplus$ and $\otimes$) denote the addition and multiplication laws of a ring, respectively

→ $(R, +, \times)$ or simply R is used for denoting an arbitrary ring

$\rightarrow R \times R'$ denotes the direct product of the rings R and R' (same notation as for groups)

$\rightarrow$ charact(R) stands for the characteristic of the ring R

$\rightarrow \mathbb{Z}$ is the ring of integers

$\rightarrow \mathbb{Z}_d$ is the ring of integers (or ring of residues) modulo d ($d \in \mathbb{N}_1$)

$\rightarrow d\mathbb{Z}$ is the ring of integers divisible by d ($d \geq 2$)

$\rightarrow \mathbb{GR}(p^s, m)$ denotes the Galois ring of characteristic p^s with p^{sm} elements (p prime and $s, m \in \mathbb{N}_1$)

$\rightarrow \mathbb{GR}(p^s, m)^*$ stands for the set of units of $\mathbb{GR}(p^s, m)$

$\rightarrow R[\xi]$ denotes the ring of polynomials in the indeterminate ξ with coefficients in the ring R

$\rightarrow \mathbb{GR}(p^s, m)[\xi]$ denotes the ring of polynomials in the indeterminate ξ with coefficients in the ring $\mathbb{GR}(p^s, m)$

$\rightarrow (\mathbb{GR}(p^s, m), +)$ is the additive group spanned by the elements of the Galois ring $\mathbb{GR}(p^s, m)$

$\rightarrow (\mathbb{GR}(p^s, m)^*, \times)$ is the multiplicative group spanned by the units of the Galois ring $\mathbb{GR}(p^s, m)$

$\rightarrow \langle a \rangle$ denotes a principal ideal spanned by the element a of a finite ring

$\rightarrow$ Tr(a) denotes the trace of the element a of a ring

$\rightarrow \phi$ denotes the generalized Frobenius automorphism for a ring

$\rightarrow \chi_b$ is an additive character of the group $(\mathbb{GR}(p^s, m), +)$ (χ_0 is the trivial additive character)

$\rightarrow \psi_k$ is a multiplicative character of the group $(\mathbb{GR}(p^s, m)^*, \times)$ (ψ_0 is the trivial multiplicative character)

$\rightarrow G_m(\psi_k, \chi_b)$ denotes the Gaussian sum for a Galois ring

Fields

→ $(\mathbb{K}, +, \times)$ or simply $\mathbb{K}$ is used for denoting an arbitrary field

→ $\mathbb{Q}$ is the field of rational numbers

→ $\mathbb{R}$ is the field of real numbers

→ $\mathbb{C}$ is the field of complex numbers

→ $\mathbb{H}$ is the field of quaternions

→ $\mathbb{GF}(p^m)$ denotes the Galois field of characteristic p with p^m elements (p prime and $m \in \mathbb{N}_1$)

→ $\mathbb{GF}(p^m)^*$ stands for the set of non-zero elements of $\mathbb{GF}(p^m)$ ($\mathbb{GF}(p^m)^* = \mathbb{GF}(p^m) \setminus \{0\}$)

→ $\mathbb{Z}_p = \mathbb{F}_p = \mathbb{GF}(p^1)$ is the field of integers (or field of residues) modulo p (p prime)

→ ${\mathbb{F}_p}^*$ stands for the set of non-zero elements of $\mathbb{F}_p$ (${\mathbb{F}_p}^* = \mathbb{F}_p \setminus \{0\}$)

→ $\mathbb{K}[\xi]$ denotes the ring of polynomials in the indeterminate ξ with coefficients in the field $\mathbb{K}$

→ $\mathbb{GF}(p^m)[\xi]$ denotes the ring of polynomials in the indeterminate ξ with coefficients in the field $\mathbb{GF}(p^m)$

→ $(\mathbb{GF}(p^m), +)$ is the additive group spanned by the elements of the Galois field $\mathbb{GF}(p^m)$

→ $(\mathbb{GF}(p^m)^*, \times)$ is the multiplicative group spanned by the non-zero elements of the Galois field $\mathbb{GF}(p^m)$

→ $\mathrm{Tr}(x)$ denotes the trace of the element x of a field

→ σ denotes the Frobenius automorphism for a field

→ χ_y is an additive character of the group $(\mathbb{GF}(p^m), +)$ (χ_0 is the trivial additive character)

$\rightarrow \psi_k$ is a multiplicative character of the group $(\mathbb{GF}(p^m)^*, \times)$ (ψ_0 is the trivial multiplicative character)

$\rightarrow \psi_{\frac{1}{2}(p^m-1)}$, with p odd, is the quadratic multiplicative character of the group $(\mathbb{GF}(p^m)^*, \times)$

$\rightarrow G_m(\psi_k, \chi_y)$ denotes the Gaussian sum for a Galois field

NOTE.– All the results that are essential to the practitioner are presented as *Proposition* or *Property* (the term proposition is used, without distinction, for lemma, theorem and proposition).

1

The Structures of Ring and Field

This chapter is devoted to some basic elements on rings and fields. The presentation is elementary and illustrated with numerous examples. The relevant references appear in the section *Mathematical literature: rings and fields* of the bibliography.

1.1. Rings

1.1.1. *The ring structure*

1.1.1.1. *Axioms of a ring*

DEFINITION 1.1.– A non-empty set R endowed with two internal composition laws (noted $+$ and $\times$) such that

– R is a commutative (or Abelian) group for the law $+$

– the law $\times$ is associative and distributive with respect to the law $+$, is a *ring*. In other words, we have

1) $\forall a \in R, \ \forall b \in R : \exists c \text{ (unique)} \in R \mid a+b=c$

2) $\forall a \in R, \ \forall b \in R, \ \forall c \in R : a+(b+c)=(a+b)+c$

3) $\exists 0 \in R \mid \forall a \in R : a+0=0+a=a$

4) $\forall a \in R, \ \exists -a \in R \mid a+(-a)=-a+a=0$

5) $\forall a \in R, \ \forall b \in R : a+b=b+a$

6) $\forall a \in R, \ \forall b \in R : \exists c \text{ (unique)} \in R \mid a \times b=c$

7) $\forall a \in R,\ \forall b \in R,\ \forall c \in R : a \times (b \times c) = (a \times b) \times c$

8) $\forall a \in R,\ \forall b \in R,\ \forall c \in R : a \times (b + c) = (a \times b) + (a \times c)$

9) $\forall a \in R,\ \forall b \in R,\ \forall c \in R : (a + b) \times c = (a \times c) + (b \times c)$

where 1 to 4 indicates that R is a group for the law $+$, 5 that this group is Abelian, 6 and 7 that the law $\times$ is an internal composition law (for 6) which is associative (for 7), and 8 and 9 that the law $\times$ is distributive (on the right for 8 and on the left for 9) with respect to the law $+$.

The element 0 is called the additive neutral element of the ring and $-a$ the additive inverse of a.

It is straightforward to show that $a \times 0 = 0 \times a = 0$ for any a in R; this follows from

$$a \times b = a \times (b + 0) \text{ and } b \times a = (b + 0) \times a$$

Consequently, $a \times (-b) = (-a) \times b = -(a \times b)$ for any a and b in the ring R; this follows from

$$a \times [b + (-b)] = 0 \text{ and } [a + (-a)] \times b = 0$$

As an immediate corollary, note that $(-a) \times (-b) = a \times b$.

1.1.1.2. *Notations*

– The element $a + (b + c) = (a + b) + c$ is simply denoted as $a + b + c$; similarly, $a \times (b \times c) = (a \times b) \times c$ is merely written as $a \times b \times c$. In the following, $a \times b$ is very often written as ab (the sign $\times$ is omitted) and $a + (-b)$ can be written as $a - b$. The multiplication has priority on the addition as in the case of ordinary addition and multiplication of real or complex numbers (although the laws $+$ and $\times$ of a ring are not necessarily those used for real or complex numbers); consequently, $a + (b \times c)$ is written $a + b \times c$ or simply $a + bc$.

– In the mathematical literature, a ring corresponding to a set R equipped with the laws $+$ and $\times$ is denoted as $(R, +, \times)$ or simply R. Here, we will often use R to denote the ring $(R, +, \times)$. When necessary, we will use $(R, +)$ and $(R, \times)$ to denote the set R endowed with the laws $+$ and $\times$, respectively. The neutral element of the Abelian group $(R, +)$ is always denoted as 0.

1.1.1.3. *Unity*

All rings possess a neutral or zero element 0 for the addition law $+$. In many mathematical and physical applications, use is made of rings with a neutral or unity or identity element, denoted as 1 in general, for the multiplication law $\times$. Such rings are then referred to as rings with unity or unitary rings (see section 1.1.9). In this regard, in addition to the nine axioms of 1.1.1.1, a tenth axiom, namely,

$$10) \quad \exists 1 \in R \mid \forall a \in R : a \times 1 = 1 \times a = a$$

is used in some books to define a ring R. In this book, we will adopt the definition involving nine axioms. This does not contradict the fact that many rings to be considered in the examples admit a unity.

For a unitary ring, it may happen that $0 = 1$. In this case, the ring contains only one element that is both the zero element and the unity element.

1.1.2. *Cardinal of a ring*

1.1.2.1. *Cardinal*

DEFINITION 1.2.– Let $(R, +, \times)$ be a ring. The cardinal of the set R, denoted as $\mathrm{Card}(R)$, is called the cardinal of the ring $(R, +, \times)$.

1.1.2.2. *Notations*

A ring $(R, +, \times)$ may contain a finite number ($\mathrm{Card}(R)$ is a finite number), a countable infinite number ($\mathrm{Card}(R)$ is then denoted $\aleph_0$) or an uncountable infinite number ($\mathrm{Card}(R)$ is then equal to $2^{\aleph_0}$) of elements. A ring that contains a finite (respectively, infinite) number of elements is called a finite (respectively, infinite) ring.

1.1.3. *Commutative ring*

1.1.3.1. *Abelian ring*

DEFINITION 1.3.– Let $(R, +, \times)$ be a ring. If the multiplication law $\times$ is commutative, i.e.

$$\forall a \in R, \ \forall b \in R : a \times b = b \times a$$

then $(R, +, \times)$ is said to be a *commutative ring* (or *Abelian ring*).

1.1.3.2. *Remark*

The addition law $+$ of a ring is always commutative. It is only when the multiplicative law $\times$ is commutative that the ring is said to be commutative.

1.1.4. *Homomorphism and isomorphism of rings*

1.1.4.1. *Homomorphism and isomorphism*

DEFINITION 1.4.– Let $(R, +, \times)$ and $(R', \oplus, \otimes)$ be two rings. A map $f : R \rightarrow R'$ such that

$$\forall a \in R, \forall b \in R : f(a + b) = f(a) \oplus f(b), \;\; f(a \times b) = f(a) \otimes f(b)$$

is called a homomorphism from $(R, +, \times)$ into $(R', \oplus, \otimes)$. If the sets R and R' have the same cardinal and the map $f : R \rightarrow R'$ is one-to-one, then f is called an isomorphism from $(R, +, \times)$ onto $(R', \oplus, \otimes)$, and the rings $(R, +, \times)$ and $(R', \oplus, \otimes)$ are said to be isomorphic.

1.1.4.2. *Remarks*

In view of the latter definition, a homomorphism of rings preserves the addition and multiplication laws of rings. Of course, this remark applies to an isomorphism. Furthermore, note that

$$f(0_R) = 0_{R'}, \quad \forall a \in R : f(-a) = -f(a)$$

where 0_R and $0_{R'}$ stand for the zero elements of $(R, +, \times)$ and $(R', \oplus, \otimes)$, respectively.

1.1.5. *Examples of rings*

1.1.5.1. *Example: the ring of integers*

Let $\mathbb{Z}$ be the infinite set of integers. The set $\mathbb{Z}$ shows a structure of ring with respect to the usual addition $+$ and the usual multiplication $\times$ of integers. The ring $(\mathbb{Z}, +, \times)$, or simply $\mathbb{Z}$, is commutative with a countable infinite number of elements ($\mathrm{Card}(\mathbb{Z}) = \aleph_0$). It is a ring with unity called the ring of integers.

1.1.5.2. *Example: the Gauss ring*

Let $\mathbb{Z}(\mathrm{i})$ be the infinite set $\{a + \mathrm{i}b \mid a, b \in \mathbb{Z}\}$ where i is the pure imaginary. The set $\mathbb{Z}(\mathrm{i})$ endowed with the addition $+$ and the multiplication $\times$ of complex numbers is a commutative ring with unity called the Gauss ring. The ring $\mathbb{Z}(\mathrm{i})$ has a countable infinite number of elements ($\mathrm{Card}(\mathbb{Z}(\mathrm{i})) = \aleph_0$).

1.1.5.3. *Example: the ring $\mathbb{Z}_4$*

Let $\mathbb{Z}_4 = \{0, 1, 2, 3\}$. One can check at once that $(\mathbb{Z}_4, +, \times)$, where $+$ and $\times$ stand for the addition and multiplication modulo 4 respectively, is a commutative ring with unity of cardinal 4. (In general, $a + b$ modulo d with $d \in \mathbb{N}_1$ is equal to the rest of the division of $a+b$ by d. Similarly, $a \times b$ modulo d with $d \in \mathbb{N}_1$ is equal to the rest of the division of $a \times b$ by d.) For more details, see Table 1.1 for the addition law and Table 1.2 for the multiplication law of the ring $\mathbb{Z}_4$.

+	0	1	2	3
0	0	1	2	3
1	1	2	3	0
2	2	3	0	1
3	3	0	1	2

Table 1.1. *Addition table for the ring $\mathbb{Z}_4$; the element at the intersection of the line a and the column b is $a + b$ (there is a symmetry with respect to the diagonal of the table, symmetry that follows from $a + b = b + a$ since any ring is an Abelian group for the law $+$)*

×	0	1	2	3
0	0	0	0	0
1	0	1	2	3
2	0	2	0	2
3	0	3	2	1

Table 1.2. *Multiplication table for the ring $\mathbb{Z}_4$; the element at the intersection of the line a and the column b is $a \times b$ (there is a symmetry with respect to the diagonal of the table, symmetry that follows from $a \times b = b \times a$ since $\mathbb{Z}_4$ is a commutative ring); $\mathbb{Z}_4$ is a ring with unity (1 is the unity element)*

1.1.5.4. *Example: two other rings of cardinal 4*

Let $A = \{0, 1, \alpha, 1 + \alpha\}$ be an abstract set endowed with the addition law $+$ and the multiplication law $\times$ given by Tables 1.3 and 1.4, respectively. We easily verify that $(A, +, \times)$ is a commutative ring with unity of cardinal 4.

$+$	0	1	α	$1+\alpha$
0	0	1	α	$1+\alpha$
1	1	0	$1+\alpha$	α
α	α	$1+\alpha$	0	1
$1+\alpha$	$1+\alpha$	α	1	0

Table 1.3. *Addition table for the ring A denoted as $\mathbb{Z}_2[\xi]/\langle\xi^2\rangle$ too*

$\times$	0	1	α	$1+\alpha$
0	0	0	0	0
1	0	1	α	$1+\alpha$
α	0	α	0	α
$1+\alpha$	0	$1+\alpha$	α	1

Table 1.4. *Multiplication table for the ring A denoted as $\mathbb{Z}_2[\xi]/\langle\xi^2\rangle$ too*

Furthermore, let us consider the abstract set $B = \{0, 1, \beta, 1+\beta\}$ equipped with the addition law $+$ and the multiplication law $\times$ defined by Tables 1.5 and 1.6, respectively. We can check that $(B, +, \times)$ is another commutative ring with unity of cardinal 4.

$+$	0	1	β	$1+\beta$
0	0	1	β	$1+\beta$
1	1	0	$1+\beta$	β
β	β	$1+\beta$	0	1
$1+\beta$	$1+\beta$	β	1	0

Table 1.5. *Addition table for the ring B denoted as $\mathbb{Z}_2[\xi]/\langle\xi+\xi^2\rangle$ too*

For reasons to be clarified in Chapter 3, the rings $(A, +, \times)$, or simply A and $(B, +, \times)$, or simply B in this example, are denoted as $\mathbb{Z}_2[\xi]/\langle\xi^2\rangle$ and $\mathbb{Z}_2[\xi]/\langle\xi+\xi^2\rangle$, respectively.

It should be observed that the rings $\mathbb{Z}_2[\xi]/\langle\xi^2\rangle$ and $\mathbb{Z}_2[\xi]/\langle\xi+\xi^2\rangle$ have the same addition table (up to the correspondence $\alpha \leftrightarrow \beta$) but different multiplication tables. Thus, they are not isomorphic. They also differ from the ring $\mathbb{Z}_4$, another ring of cardinal 4. This shows that in general, there are several different (i.e. not isomorphic) finite rings with the same cardinal.

$\times$	0	1	β	$1+\beta$
0	0	0	0	0
1	0	1	β	$1+\beta$
β	0	β	β	0
$1+\beta$	0	$1+\beta$	0	$1+\beta$

Table 1.6. *Multiplication table for the ring B denoted as $\mathbb{Z}_2[\xi]/\langle \xi + \xi^2 \rangle$ too*

1.1.5.5. *Example: the ring $\mathbb{Z}_6$*

Let $\mathbb{Z}_6 = \{0, 1, 2, 3, 4, 5\}$. The set $\mathbb{Z}_6$ endowed with the addition $+$ and the multiplication $\times$ modulo 6 is a commutative ring with unity of cardinal 6. For more details, see Tables 1.7 and 1.8 for the addition table and the multiplication table of $\mathbb{Z}_6$, respectively.

+	0	1	2	3	4	5
0	0	1	2	3	4	5
1	1	2	3	4	5	0
2	2	3	4	5	0	1
3	3	4	5	0	1	2
4	4	5	0	1	2	3
5	5	0	1	2	3	4

Table 1.7. *Addition table for the ring $\mathbb{Z}_6$*

$\times$	0	1	2	3	4	5
0	0	0	0	0	0	0
1	0	1	2	3	4	5
2	0	2	4	0	2	4
3	0	3	0	3	0	3
4	0	4	2	0	4	2
5	0	5	4	3	2	1

Table 1.8. *Multiplication table for the ring $\mathbb{Z}_6$*

1.1.5.6. *Example: the ring $\mathbb{Z}_d$*

As a generalization of examples 1.1.5.3 and 1.1.5.5, the set

$$\mathbb{Z}_d = \{0, 1, \cdots, d-1\}$$

with $d \geq 1$ (i.e. the set of remainders on division by d of all the integers, also called the set of residues modulo d) shows a structure of ring with respect to the addition $+$ and the multiplication $\times$ modulo d. Indeed, $\mathbb{Z}_d$ endowed with the law $+$ is an Abelian group for which the additive neutral or zero element is 0; every element x of $\mathbb{Z}_d$ has an inverse $d - x$ modulo d for the addition law. The multiplication law is associative, distributive with respect to the addition law and commutative; the element 1 is the multiplicative neutral or unity or identity element for the law $\times$ (for $d = 1$, the ring $\mathbb{Z}_1$ has only one element, i.e. 0, that may be considered both as an additive and a multiplicative neutral element); in general, $\mathbb{Z}_d \setminus \{0\}$ is not a group for the multiplication law because it may happen that there are some elements that do not admit an inverse for the law $\times$ (for example, 2 does not have an inverse in $\mathbb{Z}_4$ for the law $\times$). The ring $(\mathbb{Z}_d, +, \times)$ or simply $\mathbb{Z}_d$, also denoted as $\mathbb{Z}/d\mathbb{Z}$, is called the (residue class) ring of integers modulo d. It is a commutative ring, with unity, of cardinal d.

At this stage, let us remember what we call a residue class of an integer. For given $a \in \mathbb{Z}$ and $d \in \mathbb{N}_1$, the set

$$\mathcal{C}_a = \{a + dk \mid k \in \mathbb{Z}\} = \{a \bmod d\}$$

(noted $d\mathbb{Z}+a$ too) is called the Gaussian residue class of a modulo d. For given d, we have d non-equivalent residue classes corresponding to $a = 0, 1, \cdots, d-1$. These d residue classes are equivalence classes and thus constitute a partition of $\mathbb{Z}$, in the sense that $\mathbb{Z}$ is a disjoint union of the classes $\mathcal{C}_a$

$$\mathbb{Z} = \mathcal{C}_0 \cup \mathcal{C}_1 \cup \cdots \cup \mathcal{C}_{d-1} = d\mathbb{Z} \cup d\mathbb{Z} + 1 \cup \cdots \cup d\mathbb{Z} + d - 1$$

As a trivial example, for $d = 2$, the class $\mathcal{C}_0$ represents the even integers and the class $\mathcal{C}_1$ the odd integers. For d arbitrary, it is clear that the set $\{\mathcal{C}_0, \mathcal{C}_1, \cdots, \mathcal{C}_{d-1}\}$ is a ring with respect to the addition and the multiplication modulo d. We usually identify $\mathcal{C}_a$ with a so that the ring

$$\{\mathcal{C}_0, \mathcal{C}_1, \cdots, \mathcal{C}_{d-1}\} = \{d\mathbb{Z}, d\mathbb{Z} + 1, \cdots, d\mathbb{Z} + d - 1\}$$

is identified with $\mathbb{Z}_d = \{0, 1, \cdots, d-1\}$.

Note that the sum of the elements of the ring $\mathbb{Z}_d$ is equal to 0 modulo d when d is an odd (prime or not) number.

In section 1.2.5.4, we will see that when d is a prime number p (p even or odd), then the ring $\mathbb{Z}_p$ has very special properties. In this case, the particular ring $\mathbb{Z}_p$ is a field.

1.1.5.7. *Example: rings of polynomials*

– The set of polynomials $\{a_0 + a_1\xi + \cdots + a_n\xi^n,\ n \in \mathbb{N}_0\}$ in one indeterminate ξ and with coefficients $a_0, a_1, \cdots, a_n$ in $\mathbb{R}$ or $\mathbb{C}$, endowed with the addition and multiplication of polynomials, is a commutative ring with unity. This ring is denoted as $\mathbb{R}[\xi]$ or $\mathbb{C}[\xi]$ according to whether the coefficients are taken in $\mathbb{R}$ or $\mathbb{C}$. Obviously, this set has an uncountable infinite number of elements ($\mathrm{Card}(\mathbb{R}[\xi]) = \mathrm{Card}(\mathbb{C}[\xi]) = 2^{\aleph_0}$).

– As another example, especially useful in the following, let us consider a finite ring R. The set of polynomials in one indeterminate ξ, with coefficients in R, endowed with the addition and multiplication of polynomials (the addition and multiplication of coefficients is performed in R), is a ring denoted as $R[\xi]$ and referred to as the ring of polynomials over R. The elements of $R[\xi]$ are $a_0 + a_1\xi + \cdots + a_n\xi^n$ where $a_0, a_1, \cdots, a_n$ are in R and n in $\mathbb{N}_0$ (ξ is not an element of R). If R possesses a unity, then $R[\xi]$ has a unity too. The ring $R[\xi]$ is commutative if and only if R is commutative. It has a countable infinite number of elements ($\mathrm{Card}(R[\xi]) = \aleph_0$).

1.1.5.8. *Example: rings of square matrices*

– The set of square matrices of dimension $m \geq 2$, with elements in $\mathbb{R}$ or $\mathbb{C}$, endowed with the addition and multiplication of matrices is a non-commutative ring with unity and with an uncountable infinite number of elements.

– The set of square matrices of dimension $m \geq 2$, with elements in a finite ring of cardinal d, endowed with the addition and multiplication of matrices is a non-commutative ring of cardinal d^{m^2}.

1.1.5.9. *Example: a ring of triangular matrices*

Under the addition and multiplication of matrices, the set of triangular matrices of dimension 2

$$\left\{ \begin{pmatrix} a & b \\ 0 & c \end{pmatrix} \mid a, b, c \in \mathbb{Z}_d \right\}$$

is a non-commutative finite ring with unity of cardinal d^3. For further reference, let us denote this ring by M_d.

1.1.5.10. *Example: ring of even integers*

The even integers (i.e. the integers divisible by 2) form a countable infinite commutative ring without unity with respect to the usual addition and multiplication laws.

1.1.5.11. *Example: another ring without unity*

The set $R = \{0, 5, 10, 15, 20\}$ equipped with the addition $+$ and multiplication $\times$ modulo 25 is a commutative ring without unity of cardinal 5. Observe that for all a and b in R, we have $a \times b = 0$.

1.1.6. *Sub-ring of a ring*

1.1.6.1. *Sub-ring*

DEFINITION 1.5.– Let $(R, +, \times)$ be a ring and S a non-empty subset of R. If S is a ring with respect to the laws $+$ and $\times$ of $(R, +, \times)$, then the triplet $(S, +, \times)$ is said to be a *sub-ring* of $(R, +, \times)$. In other words, a sub-ring S of a ring R is a sub-group of $(R, +)$ that is stable for the multiplication law. If $(S, +, \times)$ is a sub-ring of $(R, +, \times)$, then $(R, +, \times)$ is said to be a ring extension of $(S, +, \times)$.

The ring $(S, +, \times)$, where $S = \{0\}$, and the ring $(R, +, \times)$ are trivial sub-rings of $(R, +, \times)$. The sub-rings of $(R, +, \times)$ different from $(R, +, \times)$ are called proper sub-rings of $(R, +, \times)$ (the trivial sub-ring $(\{0\}, +, \times)$ is a proper sub-ring of $(R, +, \times)$).

1.1.6.2. *Example: a sub-ring of* $R[\xi]$

Let R be a commutative ring and $R[\xi]$ be the ring of polynomials over R of elements $a = a_0 + a_1\xi + \cdots + a_n\xi^n$. The ring R can be considered as a sub-ring of $R[\xi]$, the elements of R being polynomials of type $a = a_0$ $(a_1 = a_2 = \cdots = a_n = 0)$.

1.1.6.3. *Example: a sub-ring of* $\mathbb{Z}(\mathrm{i})$

The ring $\mathbb{Z}$ is a sub-ring of the Gauss ring $\mathbb{Z}(\mathrm{i})$.

1.1.6.4. *Example: a sub-ring of* $\mathbb{Z}$

For fixed d $(d = 2, 3, 4, \cdots)$, the set $\{kd \mid k \in \mathbb{Z}\}$ of integers divisible by d endowed with the ordinary addition and multiplication laws is a commutative ring. This ring, denoted as $d\mathbb{Z}$, is a sub-ring of $\mathbb{Z}$.

1.1.6.5. *Example: a sub-ring of* $\mathbb{Z}_6$

The set $\{0, 2, 4\}$ equipped with the addition and multiplication modulo 6 is a proper sub-ring of the ring $\mathbb{Z}_6$ (the element 4 behaves as a unity element for the multiplication modulo 6 in $\{0, 2, 4\}$).

1.1.6.6. *Example: a sub-ring of* $\mathbb{Z}_8$

Let us consider the ring $\mathbb{Z}_8$. It corresponds to the set $R = \{0, 1, 2, 3, 4, 5, 6, 7\}$. The subset $S = \{0, 2, 4, 6\}$ endowed with the addition $+$ and the multiplication $\times$ modulo 8 is clearly a ring. Thus, $(S, +, \times)$ is a proper sub-ring of $\mathbb{Z}_8$. The ring $(S, +, \times)$ does not possess a unity. It is neither isomorphic to $\mathbb{Z}_4$ nor to $\mathbb{Z}_2[\xi]/\langle\xi^2\rangle$ nor to $\mathbb{Z}_2[\xi]/\langle\xi + \xi^2\rangle$ (see the examples in 1.1.5.3 and 1.1.5.4). Tables 1.9 and 1.10 illustrate the addition and multiplication laws of the ring $(S, +, \times)$, respectively.

+	0	2	4	6
0	0	2	4	6
2	2	4	6	0
4	4	6	0	2
6	6	0	2	4

Table 1.9. *Addition table for the ring* $(S, +, \times)$ *with* $S = \{0, 2, 4, 6\}$ *endowed with the addition modulo 8*

$\times$	0	2	4	6
0	0	0	0	0
2	0	4	0	4
4	0	0	0	0
6	0	4	0	4

Table 1.10. *Multiplication table for the ring* $(S, +, \times)$ *with* $S = \{0, 2, 4, 6\}$ *endowed with the multiplication modulo 8*

1.1.6.7. *Example: a sub-ring of* $\mathbb{Z}_{10}$

The ring $\mathbb{Z}_{10}$ corresponds to the set $R = \{0, 1, 2, 3, 4, 5, 6, 7, 8, 9\}$ equipped with the addition and multiplication modulo 10. The subset $S = \{0, 2, 4, 6, 8\}$ of R is a unitary ring with respect to the addition $+$ and multiplication $\times$ modulo 10. Thus, $(S, +, \times)$ is a proper sub-ring of $\mathbb{Z}_{10}$. In fact, $(S, +, \times)$ is isomorphic to the ring $\mathbb{Z}_5$ (which is a field too). The isomorphism $\mathbb{Z}_5 \leftrightarrow (S, +, \times)$ is described by the following $\mathbb{Z}_5 \leftrightarrow (S, +, \times)$ correspondences

$$0 \leftrightarrow 0,\ 1 \leftrightarrow 6,\ 2 \leftrightarrow 2,\ 3 \leftrightarrow 8,\ 4 \leftrightarrow 4$$

Note that 6 is the unity element of $(S, \times)$.

1.1.7. *Ideal of a ring*

1.1.7.1. *Ideal*

DEFINITION 1.6.– Let $(R, +, \times)$ be a commutative ring. A non-empty subset I of R such that

– $(I, +)$ is a sub-group of the additive group $(R, +)$

– I is stable under the multiplication by any element of R, i.e.

$$\forall i \in I, \forall r \in R : i \times r \in I$$

is called an *ideal* of $(R, +, \times)$.

It should be noted that the axiom $(I, +)$ *is a sub-group of the additive group* $(R, +)$ could be replaced by the weaker axiom

$$\forall i \in I, \forall j \in I : i + j \in I$$

From the definition, it is easy to show that the triplet $(I, +, \times)$ is a sub-ring of R according to definitions 1.1 and 1.5 of a ring and a sub-ring, respectively (definition 1.1 does not include the axiom of unity element as is the case in some books). Thus, an ideal of a ring is necessarily a sub-ring, but a sub-ring of a ring is not necessarily an ideal (as a trivial example: $\mathbb{Z}$ is a sub-ring of the ring $\mathbb{Z}(\mathrm{i})$ but is not an ideal of $\mathbb{Z}(\mathrm{i})$). An ideal of a unitary ring R that contains the unity of R is nothing but R.

In terms of sets, we have $rI = Ir \subset I$ (remember $rI = \{r \times i \mid i \in I\} = Ir$).

Note that $\{0\}$ and R are ideals of the ring R; they are called trivial ideals. An ideal I of a ring R is said to be a *proper ideal* if $I \neq R$. The ideal $\{0\}$ of R is a proper ideal if $\mathrm{Card}(R) > 1$.

For a non-commutative ring, the preceding definition can be replaced by the definitions of a left ideal and a right ideal. An ideal of a non-commutative ring that is both a left ideal and a right ideal is simply called an ideal.

1.1.7.2. *Maximal ideal*

DEFINITION 1.7.– A proper ideal I of a ring R such that there is no proper ideal strictly containing I is called a *maximal ideal* of R.

1.1.7.3. *Principal ideal*

DEFINITION 1.8.– Let $(R, +, \times)$ be a commutative ring. For fixed a in R

$$I = aR = \{a \times x \mid x \in R\}$$

is an ideal of $(R, +, \times)$ (the elements of I consist of the different elements $a \times x$ for fixed a and x ranging). This ideal is called a *principal ideal.* It is generated by a single element a of R and denoted as $\langle a \rangle$.

There are rings for which all ideals are principal.

Note that the notions of left principal ideal and right principal ideal exist for a non-commutative ring.

1.1.7.4. *Example: the ideal $\langle 2 \rangle$ of $\mathbb{Z}_4$*

The elements 0 and 2 of the ring $\mathbb{Z}_4$ form a principal ideal of $\mathbb{Z}_4$ spanned by 2 and denoted as $\langle 2 \rangle$. Tables 1.11 and 1.12 are the addition and multiplication tables for the ideal $\langle 2 \rangle$, respectively. Observe that $\langle 2 \rangle$ is a ring that does not possess a unity. It is not isomorphic to the ring $\mathbb{Z}_2$ (which is also a field).

+	0	2
0	0	2
2	2	0

Table 1.11. *Addition table for the ring $\langle 2 \rangle$, a principal ideal of $\mathbb{Z}_4$*

×	0	2
0	0	0
2	0	0

Table 1.12. *Multiplication table for the ring $\langle 2 \rangle$, a principal ideal of $\mathbb{Z}_4$*

1.1.7.5. *Example: the ideal $\langle 2 \rangle$ of $\mathbb{Z}_8$*

The set $S = \{0, 2, 4, 6\}$ constitutes a sub-ring of $\mathbb{Z}_8$ (see 1.1.6.6). This sub-ring is a principal ideal of $\mathbb{Z}_8$ spanned by the element 2. It is denoted by $\langle 2 \rangle$ (see Tables 1.9 and 1.10 for the addition and multiplication laws of $\langle 2 \rangle$, respectively). The distinct principal ideals of $\mathbb{Z}_8$ are $\langle 0 \rangle = \{0\}$, $\langle 4 \rangle = \{0, 4\}$, $\langle 2 \rangle = \{0, 2, 4, 6\}$ and $\langle 1 \rangle = \mathbb{Z}_8$. Note that $\langle 6 \rangle = \langle 2 \rangle$ and $\langle 3 \rangle = \langle 5 \rangle = \langle 7 \rangle = \langle 1 \rangle$.

1.1.7.6. *Example: the principal ideals of* $\mathbb{Z}_9$

The ring $\mathbb{Z}_9$ admits three principal ideals, viz. $\langle 0 \rangle = \{0\}$, $\langle 1 \rangle = \mathbb{Z}_9$ and $\langle 3 \rangle = \{0, 3, 6\}$. The proper ideal $\langle 3 \rangle$ of $\mathbb{Z}_9$ is a maximal ideal.

1.1.7.7. *Example: the ideal* $\langle 5 \rangle$ *of* $\mathbb{Z}_{25}$

The set $\{0, 5, 10, 15, 20\}$ endowed with the addition and multiplication modulo 25 is a ring (see 1.1.5.11). This ring is a sub-ring of $\mathbb{Z}_{25}$. It is a principal ideal of $\mathbb{Z}_{25}$, spanned by the element 5 and denoted by $\langle 5 \rangle$.

1.1.7.8. *Example: the ideals* $2\mathbb{Z}$, $5\mathbb{Z}$ *and* $d\mathbb{Z}$ *of* $\mathbb{Z}$

The set of even integers endowed with the ordinary addition and multiplication laws is a commutative ring denoted as $2\mathbb{Z}$. It is a sub-ring of $\mathbb{Z}$. This sub-ring is a principal ideal of $\mathbb{Z}$. It is also denoted as $\langle 2 \rangle$.

The principal ideal of $\mathbb{Z}$ generated by 5 consists of the elements $5k$ where k belongs to $\mathbb{Z}$. It is a sub-ring of $\mathbb{Z}$ denoted by $5\mathbb{Z}$ or $\langle 5 \rangle$.

More generally, for d positive integer, the ring $d\mathbb{Z}$ is a principal ideal of $\mathbb{Z}$ generated by d. It is also denoted as $\langle d \rangle$. Indeed, every ideal of $\mathbb{Z}$ is a principal ideal. If $d = p$ is prime, then $p\mathbb{Z}$ is a maximal ideal of $\mathbb{Z}$.

1.1.8. *Quotient ring*

Let I be a proper ideal of a commutative ring $(R, +, \times)$. The group $(I, +)$ is a sub-group of the additive group $(R, +)$. Thus, according to group theory (see 5.2.10), we can form the quotient group R/I whose elements are the various cosets $(r + I)$ of R with respect to I (the coset $(r + I)$ is the set $\{r + i \mid i \in I\}$ with $r \in R$). Indeed, R/I can be endowed with a ring structure as defined in the following proposition.

PROPOSITION 1.1.– Let I be a proper ideal of a commutative ring $(R, +, \times)$. The set of the various cosets $(r + I)$ equipped with the addition law $\oplus$ and the multiplication law $\otimes$ defined by

$$(r + I) \oplus (s + I) = (r + s + I)$$

$$(r + I) \otimes (s + I) = (r \times s + I)$$

$(r \in R, s \in R)$ is a ring called the residue class ring of R modulo I or quotient ring of R by I and denoted as $(R/I, \oplus, \otimes)$ or simply R/I. Furthermore, the surjective ring homomorphism

$$R \to R/I$$

$$r \mapsto (r + I)$$

admits I as kernel.

1.1.9. *Unitary ring*

1.1.9.1. *Ring with unity*

DEFINITION 1.9.– Let $(R, +, \times)$ be a ring. If $(R, +, \times)$ has a neutral element, denoted as 1, in general with respect to the multiplication law $\times$, then $(R, +, \times)$ is said to be a *unitary ring*. In detail, for a unitary ring $(R, +, \times)$

$$\exists 1 \in R \mid \forall a \in R : a \times 1 = 1 \times a = a$$

The element 1 is also called unity or identity element of $(R, +, \times)$.

1.1.9.2. *Counter-example: the ring of integers divisible by* d

The commutative ring $2\mathbb{Z}$ of even integers is a ring without unity for the multiplication; it is not a unitary ring. More generally, the ring $d\mathbb{Z}$ of integers divisible by d ($d = 2, 3, 4, \cdots$) is not a unitary ring.

1.1.9.3. *Example: unitary rings* $\mathbb{Z}$ *and* $\mathbb{Z}_d$

The rings $\mathbb{Z}$ and $\mathbb{Z}_d$ are unitary rings with the integer 1 as unity for the multiplication law.

1.1.9.4. *Example: unitary ring of polynomials*

The ring of polynomials with coefficients in $\mathbb{R}$ or $\mathbb{C}$ is a unitary ring with the integer 1 as unity for the multiplication law.

1.1.9.5. *Example: unitary ring of matrices*

The ring of matrices on $\mathbb{C}$ of dimension $m \geq 2$ is a unitary non-commutative ring with the identity matrix of dimension m as unity for the multiplication law.

1.1.10. *Characteristic of a unitary ring*

1.1.10.1. *Characteristic*

DEFINITION 1.10.– Let $(R, +, \times)$, with $R \neq \{0\}$, be a unitary ring. The smallest positive integer s ($s \geq 2$) such that

$$\forall x \in R : (1 + 1 + \cdots + 1) \times x = 0 \Leftrightarrow 1 + 1 + \cdots + 1 = 0$$

where the sum contains s terms is called the *characteristic* of $(R, +, \times)$. If there is no value of s for which $1 + 1 + \cdots + 1 = 0$, the unitary ring is said to be of characteristic 0. The characteristic of the ring R is denoted as charact(R).

It is worth noting that the characteristic of a unitary ring is a non-negative integer. We will see that the characteristic of a field (i.e. a particular ring) can be 0 or a prime number.

1.1.10.2. *Examples*

The characteristic of the rings (defined in 1.1.5.3 and 1.1.5.4) $\mathbb{Z}_4$, $\mathbb{Z}_2[\xi]/\langle\xi^2\rangle$ and $\mathbb{Z}_2[\xi]/\langle\xi + \xi^2\rangle$ of cardinal 4 is 4, 2 and 2, respectively. The ring $\mathbb{Z}_d$ is of characteristic d. The ring $\mathbb{Z}$ is of characteristic 0.

1.1.10.3. *Characteristic and cardinal*

PROPOSITION 1.2.– Let R be a finite unitary ring. Then, charact$(R) \neq 0$ and charact(R) divides Card(R).

1.1.10.4. *A remarkable identity*

PROPOSITION 1.3.– Let $(R, +, \times)$ be a commutative unitary ring of characteristic p. Then, the formula

$$\forall a \in R, \ \forall b \in R : (a + b)^p \equiv a^p + b^p \bmod p$$

holds when p is prime (even or odd).

PROOF.– The Newton binomial formula can be applied to $(a + b)^p$. Then

$$(a + b)^p = \sum_{k=0}^{p} C_p^k \, a^{p-k} \times b^k$$

where

$$C_p^k = \frac{p!}{k!(p-k)!} = p\frac{(p-1)!}{k!(p-k)!} = p\frac{(p-k+1)(p-k+2)\cdots(p-1)}{1 \times 2 \times \cdots \times k}$$

Since p is prime, C_p^k is a positive integer proportional to p except for $k = 0$ and $k = p$ (p divides the binomial coefficients C_p^k for $0 < k < p$). Therefore, the sole non-vanishing elements in the sum $\sum_{k=0}^{p} C_p^k \, a^{p-k} \times b^k$ are a^p (for $k = 0$, $b^k = 1$) and b^p (for $k = p$, $a^{p-k} = 1$). □

Note that the formula $(a + b)^p \equiv a^p + b^p \bmod p$ works for any couple (a, b) of a non-commutative ring $(R, +, \times)$ satisfying $a \times b = b \times a$.

Of course, $(a + b)^d \equiv a^d + b^d \bmod d$ is not valid when the characteristic d of the ring R is a composite number (different from a prime power) rather than a prime.

1.1.11. *Unit in a unitary ring*

1.1.11.1. *Unit*

DEFINITION 1.11.– Let $(R, +, \times)$ be a unitary ring. The element u in R is called a left (or right) unit if there exists an element v in R such that $u \times v = 1$ (or $v \times u = 1$).

A left unit that is also a right unit is simply called a unit. Of course, for a commutative unitary ring, a left unit is a right unit too and *vice versa*.

One also says that a left (or right) unit is an invertible element, i.e. an element that admits a right (or left) inverse with respect to the multiplication law. More precisely, if $u \times v = 1$ (or $v \times u = 1$) then v is called a right (or left) inverse of u.

1.1.11.2. *Example: units in $\mathbb{Z}(\mathrm{i})$, $\mathbb{Z}$, $\mathbb{Z}_4$, $\mathbb{Z}_6$ and $\mathbb{Z}_{14}$*

The Gauss ring $\mathbb{Z}(\mathrm{i})$ has four units (1, i, -1 and $-$i) and the ring $\mathbb{Z}$ has two units (1 and -1). In the ring $\mathbb{Z}_4$, there are two units namely 1 and 3 (see Table 1.2). The ring $\mathbb{Z}_6$ also has two units, namely 1 and 5 (see Table 1.8). The elements 1, 3, 9, 13, 11 and 5 of the ring $\mathbb{Z}_{14}$ are units. More generally, the element a of the ring $\mathbb{Z}_d$, with d different from a prime, is a unit if a is co-prime to d.

1.1.11.3. *Group of units*

PROPOSITION 1.4.– In a unitary ring, the set of all the units form a group with respect to the multiplication law of the ring.

1.1.11.4. *Example: group of units in* $\mathbb{Z}(\mathrm{i})$, $\mathbb{Z}$, $\mathbb{Z}_4$, $\mathbb{Z}_6$ *and* $\mathbb{Z}_{14}$

The units 1, i, -1 and $-\mathrm{i}$ of the Gauss ring $\mathbb{Z}(\mathrm{i})$ constitute a group of order 4 isomorphic to the cyclic group C_4. For the ring $\mathbb{Z}$, the units 1 and -1 span a group isomorphic to the abstract group of order 2. The same result holds for the units 1 and 3 of the ring $\mathbb{Z}_4$ and for the units 1 and 5 of the ring $\mathbb{Z}_6$. The units $1, 3, 9, 13, 11$ and 5 of the ring $\mathbb{Z}_{14}$ form a group of order 6 isomorphic to the cyclic group C_6.

1.1.11.5. *Example: group of units in* M_4

The units of the ring M_4 of 2×2 triangular matrices with coefficients in $\mathbb{Z}_4$ (see 1.1.5.9 for the definition of M_d) span a non-commutative group of order 16.

1.1.12. ***Zero divisor in a ring***

1.1.12.1. *Zero divisor*

DEFINITION 1.12.– An element a of a ring $(R, +, \times)$ is a left (respectively, right) *zero divisor* if there exists b in $(R, +, \times)$, with $b \neq 0$, such that $a \times b = 0$ (respectively, $b \times a = 0$).

According to this definition, the element 0 is a left and right zero divisor (called trivial zero divisor). Of course, there is no difference between left and right zero divisors for a commutative ring. In general (for a commutative or non-commutative ring), an element that is both a left zero divisor and a right zero divisor is simply called a zero divisor.

1.1.12.2. *Example: zero divisors in* $\mathbb{Z}_4$ *and* $\mathbb{Z}_6$

The ring $\mathbb{Z}_4$ admits two zero divisors, viz. 0 and 2. The ring $\mathbb{Z}_6$ admits four zero divisors, viz. $0, 2, 3$ and 4.

1.1.12.3. *Example: zero divisors in a matrix ring*

In the ring of real square matrices of dimension 2, we have

$$M = \begin{pmatrix} 2 & 1 \\ 0 & 0 \end{pmatrix}, N = \begin{pmatrix} 1 & -2 \\ -2 & 4 \end{pmatrix} \Rightarrow MN = \begin{pmatrix} 0 & 8 \\ 0 & 9 \end{pmatrix} M = N \begin{pmatrix} 2 & 2 \\ 1 & 1 \end{pmatrix} = O$$

so that M and N are zero divisors (here O stands for the null matrix of dimension 2).

1.1.12.4. *Mutual exclusion*

PROPOSITION 1.5.– Every element of a finite unitary ring is either a unit or a zero divisor.

In other words, a finite unitary ring contains zero divisors and units only. The proposition does not hold for infinite unitary rings: the field $\mathbb{Z}$ contains one zero divisor (0), two units (1 and -1) and an enumerable infinity of elements that are neither zero divisors nor units.

Note that a proper ideal of a unitary ring R does not contain the unity of R. A proper ideal of a finite unitary ring does not contain units (it contains only zero divisors).

1.1.12.5. *Example: $\mathbb{Z}_4$, $\mathbb{Z}_6$ and $\mathbb{Z}_{10}$*

The ring $\mathbb{Z}_4$ has two units (1 and 3) and two zero divisors (0 and 2). The ring $\mathbb{Z}_6$ has two units (1 and 5) and four zero divisors (0, 2, 3 and 4). The ring $\mathbb{Z}_{10}$ has four units (1, 3, 7 and 9) and six zero divisors (0, 2, 4, 5, 6 and 8).

1.1.12.6. *Example: units in $\mathbb{Z}_{p^s}$, p prime and s positive integer*

The ring $\mathbb{Z}_9 = \mathbb{Z}_{3^2}$ contains three zero divisors (0, 3 and 6) and six units (1, 2, 4, 5, 7 and 8). Observe that each unit of $\mathbb{Z}_{3^2}$ is co-prime to $p = 3$. More generally, all elements of $\mathbb{Z}_{p^s} = \{0, 1, \cdots, p^s - 1\}$, p prime and s positive integer, co-prime to p are units and the other elements are zero divisors.

1.1.12.7. *Example: $\mathbb{Z}_2[\xi]/\langle\xi^2\rangle$ and $\mathbb{Z}_2[\xi]/\langle\xi + \xi^2\rangle$*

The ring $\mathbb{Z}_2[\xi]/\langle\xi^2\rangle$ in 1.1.5.4 has two units (1 and $1 + \alpha$) and two zero divisors (0 and α). The ring $\mathbb{Z}_2[\xi]/\langle\xi + \xi^2\rangle$ in 1.1.5.4 has one unit (1) and three zero divisors (0, β and $1 + \beta$).

1.1.12.8. *Example: M_4*

The $4^3 = 64$ elements of the ring M_4 of 2×2 triangular matrices with coefficients in $\mathbb{Z}_4$ (see 1.1.5.9 for the definition of M_d) can be separated into 16 units (see 1.1.11.5) and 48 zero divisors.

1.1.13. *Integrity ring*

1.1.13.1. *Ring with only one zero divisor*

DEFINITION 1.13.– A ring is called an *integrity ring* if it does not admit zero divisor except 0 (the trivial zero divisor).

1.1.13.2. *Example: the ring $\mathbb{Z}$*

The ring $\mathbb{Z}$ is an integrity ring.

1.1.13.3. *Solutions of $a \times b = 0$ in an integrity ring*

PROPOSITION 1.6.– In an integrity ring $(R, +, \times)$, we have

$$a \times b = 0 \Leftrightarrow a = 0 \text{ or } b = 0, \quad a \neq 0 \text{ and } b \neq 0 \Leftrightarrow a \times b \neq 0$$

with a and b in $(R, +, \times)$.

1.2. Fields

1.2.1. *The field structure*

1.2.1.1. *Axioms of a field*

DEFINITION 1.14.– A non-empty set $\mathbb{K}$ endowed with two internal composition laws, denoted as $+$ with 0 as the neutral element and $\times$ with 1 as the neutral element, such that

– $\mathbb{K}$ is a commutative (or Abelian) group for the law $+$

– $\mathbb{K}^* = \mathbb{K} \setminus \{0\}$ is a group for the law $\times$

– the law $\times$ is distributive with respect to the law $+$ is a *field*. In other words, we have

1) $\forall a \in \mathbb{K},\ \forall b \in \mathbb{K} : \exists c \text{ (unique)} \in \mathbb{K} \mid a + b = c$

2) $\forall a \in \mathbb{K},\ \forall b \in \mathbb{K},\ \forall c \in \mathbb{K} : a + (b + c) = (a + b) + c$

3) $\exists 0 \in \mathbb{K} \mid \forall a \in \mathbb{K} : a + 0 = 0 + a = a$

4) $\forall a \in \mathbb{K},\ \exists -a \in \mathbb{K} \mid a + (-a) = -a + a = 0$

5) $\forall a \in \mathbb{K},\ \forall b \in \mathbb{K} : a + b = b + a$

6) $\forall a \in \mathbb{K},\ \forall b \in \mathbb{K} : \exists c\,(\text{unique}) \in \mathbb{K} \mid a \times b = c$

7) $\forall a \in \mathbb{K},\ \forall b \in \mathbb{K},\ \forall c \in \mathbb{K} : a \times (b \times c) = (a \times b) \times c$

8) $\exists 1 \in \mathbb{K} \mid \forall a \in \mathbb{K} : a \times 1 = 1 \times a = a$

9) $\forall a \in \mathbb{K}^*,\ \exists a^{-1} \in \mathbb{K} \mid a \times a^{-1} = a^{-1} \times a = 1$

10) $\forall a \in \mathbb{K},\ \forall b \in \mathbb{K},\ \forall c \in \mathbb{K} : a \times (b + c) = (a \times b) + (a \times c)$

11) $\forall a \in \mathbb{K},\ \forall b \in \mathbb{K},\ \forall c \in \mathbb{K} : (a + b) \times c = (a \times c) + (b \times c)$

where 1 to 4 indicates that $\mathbb{K}$ is a group for the addition law $+$, 5 that this group is Abelian, 6 to 9 that $\mathbb{K}^*$ is a group for the multiplication law $\times$, and 10 and 11 that the law $\times$ is distributive (on the right for 10 and on the left for 11) with respect to the law $+$.

Axioms 1 to 7 plus 10 and 11 are the axioms of a ring. Thus, the structure of field covers the structure of ring. Each field is a ring (a field is a particular ring). Indeed, a field is a unitary ring for which every non-zero element admits an inverse for the multiplication law $\times$ (a field does not have zero divisor except the trivial zero divisor 0: every non-zero element is a unit). On the contrary, a ring is not necessarily a field. Note that every finite integrity ring with unity is a field; however, an infinite integrity ring with unity is not necessarily a field: $\mathbb{Z}$ is an infinite integrity ring with unity but is not a field.

A field $\mathbb{K}$ has only two ideals, viz. the trivial ideals $\{0\}$ and $\mathbb{K}$ ($\{0\} = \langle 0 \rangle$ and $\mathbb{K} = \langle 1 \rangle$ are principal ideals). Indeed, a ring R is a field if and only if the sole ideals of R are $\{0\}$ and R.

In several textbooks (mainly in English/American literature), the multiplication law of a field is supposed to be commutative. In such textbooks, axioms 1 to 11 define a *division ring* or *skewfield*. In the finite case (which is the main case to be treated in the following chapters), there is no difference between a field and a division ring because axioms 1 to 11 imply that the multiplication is commutative in this case: every finite division ring is a field (see *Wedderburn's theorem* in Chapter 2). We adopt here axioms 1 to 11 for defining a field in general.

Note that a field possesses at least two elements (0 and 1 with $1 \neq 0$). The additive neutral element (or zero element) and the multiplicative neutral element (also called unity element or identity element) of a field are always denoted by 0 and 1, respectively. The elements $-a$ and a^{-1} are called the additive inverse (with respect to the law $+$) and the multiplicative inverse (with respect to the law $\times$) of a, respectively.

1.2.1.2. *Notations*

– The element $a + (b + c) = (a + b) + c$ is simply denoted as $a + b + c$; similarly, $a \times (b \times c) = (a \times b) \times c$ is merely written as $a \times b \times c$. In an expression involving the signs $+$ and $\times$, priority is given to the sign $\times$; for instance, $a \times b + c \times d$ means $(a \times b) + (c \times d)$. In the following, $a \times b$ is very often written as ab (the sign $\times$ is omitted).

– In the mathematical literature, a field corresponding to a set $\mathbb{K}$ equipped with the laws $+$ and $\times$ is denoted by the triplet $(\mathbb{K}, +, \times)$; in this book, we will use simply $\mathbb{K}$ when there is no ambiguity. Finally, we will use $(\mathbb{K}, +)$ and $(\mathbb{K}^*, \times)$ to denote the sets $\mathbb{K}$ and $\mathbb{K}^* = \mathbb{K} \setminus \{0\}$ endowed with the laws $+$ and $\times$, respectively. Note that $(\mathbb{K}, +)$ is a commutative group, but that $(\mathbb{K}^*, \times)$ is not necessarily a commutative group.

1.2.1.3. *Product of two non-zero elements*

PROPOSITION 1.7.– In a field $(\mathbb{K}, +, \times)$, the product of two non-zero elements is different from 0, that is to say

$$a \neq 0, \quad b \neq 0 \Rightarrow a \times b \neq 0$$

for a and b in $\mathbb{K}^*$.

PROOF.– The proof easily follows from the fact that, for $a \neq 0$ and $b \neq 0$, both elements a and b admit a multiplicative inverse. Thus, a field has no zero divisor (except 0, the trivial divisor). In other words, if $a \times b = 0$, then $a = 0$ or $b = 0$ (or $a = b = 0$). □

1.2.2. *Cardinal of a field*

1.2.2.1. *Number of elements of a field*

DEFINITION 1.15.– Let $(\mathbb{K}, +, \times)$ be a field. The cardinal of the set $\mathbb{K}$, denoted as $\text{Card}(\mathbb{K})$, is called the cardinal of the field $(\mathbb{K}, +, \times)$.

A field may have a countable infinite number of elements (the cardinal of the field is $\aleph_0$), an uncountable infinite number of elements (the cardinal of the field is $2^{\aleph_0}$) or a finite number of elements.

1.2.2.2. *Remark*

In the following, we will be mainly concerned with finite fields (i.e. fields with a finite number of elements). Therefore, most of the examples will be for finite fields. Nevertheless, we will also give some examples concerning infinite fields (i.e. fields with an infinite number of elements).

1.2.3. *Commutative field*

1.2.3.1. *Field with a commutative multiplication*

DEFINITION 1.16.– A field $(\mathbb{K}, +, \times)$ is said to be commutative if the multiplication law $\times$ is commutative.

1.2.3.2. *Remark*

As already noted in many books, a field is supposed to be equipped with a commutative multiplication. This is not the case in this monograph: an infinite field can be commutative or non-commutative, but all finite fields are necessarily commutative, a result known as Wedderburn's theorem (see Chapter 2).

1.2.4. *Isomorphism and automorphism of fields*

The definitions of homomorphism and isomorphism of rings apply to fields since a field is a particular ring. The definition of an isomorphism of fields can be precised as follows.

1.2.4.1. *Isomorphism*

DEFINITION 1.17.– Let $(\mathbb{K}, +, \times)$ and $(\mathbb{K}', \oplus, \otimes)$ be two fields of the same cardinal. A one-to-one map $f : \mathbb{K} \to \mathbb{K}'$ such that

$$\forall x \in \mathbb{K}, \forall y \in \mathbb{K} : f(x+y) = f(x) \oplus f(y), \ f(x \times y) = f(x) \otimes f(y)$$

is called an isomorphism from $(\mathbb{K}, +, \times)$ onto $(\mathbb{K}', \oplus, \otimes)$. The fields $(\mathbb{K}, +, \times)$ and $(\mathbb{K}', \oplus, \otimes)$ are said to be isomorphic.

Note that

$$f(0_{\mathbb{K}}) = 0_{\mathbb{K}'}, \quad f(1_{\mathbb{K}}) = 1_{\mathbb{K}'}$$

so that the images $f(0_{\mathbb{K}})$ and $f(1_{\mathbb{K}})$ are the zero element and the unity element of $\mathbb{K}'$, respectively ($0_{\mathbb{K}}$ and $1_{\mathbb{K}}$ stand for the zero element and the unity element of $\mathbb{K}$, respectively). Note also that

$$f(-x) = -f(x), \quad f(x^{-1}) = f(x)^{-1}$$

for any x in $\mathbb{K}^*$.

1.2.4.2. *Automorphism*

DEFINITION 1.18.– If $\mathbb{K}$ and $\mathbb{K}'$ are identical, then the isomorphism is called an automorphism.

1.2.5. *Examples of fields*

1.2.5.1. *Counter-example:* $(\mathbb{Z}_4, +, \times)$ *is not a field*

Although the ring $(\mathbb{Z}_4, +, \times)$ is a unitary ring (it admits a unity element for the multiplication law $\times$), $(\mathbb{Z}_4^*, \times)$ is not a group. As a matter of fact, from Table 1.2, we have $2 \times 2 = 0 \notin \mathbb{Z}_4^*$. Another way to see that $(\mathbb{Z}_4^*, \times)$ is not a group is to consider $2 \times 1 = 2 \times 3$ which shows that if 2 had an inverse, we should have $1 = 3$. Therefore, $(\mathbb{Z}_4, +, \times)$ is a ring but not a field.

1.2.5.2. *Counter-example:* $(\mathbb{Z}, +, \times)$ *is not a field*

The ring $(\mathbb{Z}, +, \times)$ admits a unity element with respect to the law $\times$ (the ring $(\mathbb{Z}, +, \times)$ is a unitary ring). It is also an integrity ring. However, an element of $\mathbb{Z}$ has no inverse in general with respect to the law $\times$ (except the elements 1 and -1 which are their own inverse). Therefore, $(\mathbb{Z}, +, \times)$ is a ring but not a field.

1.2.5.3. *Example: the field $\mathbb{Z}_5$*

Let $\mathbb{Z}_5 = \{0, 1, 2, 3, 4\}$. The set $\mathbb{Z}_5$ endowed with the addition $+$ and the multiplication $\times$ modulo 5 is a commutative field of cardinal 5. For more details, see Tables 1.13 and 1.14 for the addition and multiplication tables of $\mathbb{Z}_5$, respectively. Observe that the non-zero elements of $\mathbb{Z}_5$ are given by powers of 2: $2^4 \equiv 1 \bmod 5$, $2^1 = 2$, $2^3 \equiv 3 \bmod 5$, $2^2 = 4$. In this respect, we say that 2 is a primitive element of $\mathbb{Z}_5$. Similarly, 3 is a primitive element of $\mathbb{Z}_5$ in the sense that every non-zero element of $\mathbb{Z}_5$ is a power of 3. Note that 2 is the multiplicative inverse of 3 and reciprocally.

+	0	1	2	3	4
0	0	1	2	3	4
1	1	2	3	4	0
2	2	3	4	0	1
3	3	4	0	1	2
4	4	0	1	2	3

Table 1.13. *Addition table for the field $\mathbb{Z}_5$ (observe that this table is also the group table of the cyclic group C_5)*

×	0	1	2	3	4
0	0	0	0	0	0
1	0	1	2	3	4
2	0	2	4	1	3
3	0	3	1	4	2
4	0	4	3	2	1

Table 1.14. *Multiplication table for the field $\mathbb{Z}_5$ (observe that the part of the table corresponding to the elements $1 \equiv 2^4$ mod 5, $2 = 2^1$, $4 = 2^2$ and $3 \equiv 2^3$ mod 5 is also the group table of the cyclic group C_4)*

1.2.5.4. *Example: the field $\mathbb{Z}_p$, with p prime*

More generally, let us consider the set $\mathbb{Z}_p = \{0, 1, \cdots, p-1\}$ where p is a prime number ($p \geq 2$). One can check that the set $\mathbb{Z}_p$ endowed with the addition $+$ and multiplication $\times$ modulo p is a commutative field of cardinal p. This field is denoted as $(\mathbb{Z}_p, +, \times)$ or $\mathbb{Z}_p$ or $\mathbb{Z}/p\mathbb{Z}$ or $\mathbb{F}_p$ and is called the field of residue classes of integers modulo p or, simply, the field of integers modulo p. In the following, we will mainly use the notation $\mathbb{F}_p$. For $p = 2$, the field $\mathbb{F}_2$ is called a binary field.

It should be realized that $\mathbb{Z}_d$, with d different from a prime number, is a ring but not a field. This can be seen as follows. Indeed, in $\mathbb{Z}_d$, with d not prime, it is possible to find elements different from 0 for which the product is 0. For example, the product $(d-m) \times (d-n)$ of the elements $d-m$ and $d-n$, with $m \in \{1, 2, \cdots, d-1\}$ and $n \in \{1, 2, \cdots, d-1\}$, is 0 when d is a composite number equal to mn modulo d. Thus $\mathbb{Z}_d$, with d different from a prime number, is a unitary commutative ring; it is not an integrity ring.

To sum up, the ring $\mathbb{Z}_d$, denoted as $\mathbb{Z}/d\mathbb{Z}$ too, of residues modulo d is a field if and only if d is a prime number. The ring $\mathbb{Z}/d\mathbb{Z}$ for d, different from a prime number, or the field $\mathbb{Z}/p\mathbb{Z}$ for p, a prime number, is the quotient of $\mathbb{Z}$ by the congruence relation modulo d or p, respectively.

1.2.5.5. *Example: the field of rational numbers*

The set $\mathbb{Q}$ of rational numbers is a commutative field, with a countable infinite number of elements, with respect to the usual addition and multiplication.

1.2.5.6. *Example: the field $\mathbb{Q}(\sqrt{p})$, with p prime*

The set

$$\mathbb{Q}(\sqrt{p}) = \{a + b\sqrt{p} \mid a, b \in \mathbb{Q}\}$$

where p is a prime number (even: $p = 2$ or odd: $p = 3, 5, 7, \cdots$) is a field with respect to the usual addition and multiplication laws of real numbers. It is a commutative field with a countable infinite number of elements. Note that the inverse (with respect to the multiplication law) of $a + b\sqrt{p}$ is $a' + b'\sqrt{p}$ where

$$a' = \frac{a}{a^2 - pb^2}, \qquad b' = -\frac{b}{a^2 - pb^2}$$

in the cases $(a = 0,\ b \neq 0)$, $(a \neq 0,\ b = 0)$ and $ab \neq 0$ (if $ab \neq 0$, the denominator $a^2 - pb^2$ is different from zero since $a, b \in \mathbb{Q}$ and p is prime). By anticipating the developments in Chapter 2, we can denote $\mathbb{Q}(\sqrt{p})$ as $\mathbb{Q}[\xi]/\langle -p + \xi^2 \rangle$.

1.2.5.7. *Example: the field* $\mathbb{Q}(\sqrt{2}, \sqrt{3})$

The set

$$\mathbb{Q}(\sqrt{2}, \sqrt{3}) = \left\{ x = a + b\sqrt{2} + c\sqrt{3} + d\sqrt{2}\sqrt{3} \mid a, b, c, d \in \mathbb{Q} \right\}$$

is a field with respect to the usual addition and multiplication laws of real numbers. It is a commutative field with a countable infinite number of elements.

This example is appropriate for introducing the notion (to be precised in Chapter 2 in the case of a field with a finite number of elements) of a primitive element of a field. Here, we note that

$$\sqrt{2} = \frac{1}{2}\alpha(\alpha^2 - 9), \quad \sqrt{3} = \frac{1}{2}\alpha(11 - \alpha^2)$$

where

$$\alpha = \sqrt{2} + \sqrt{3}$$

Therefore, every element x of $\mathbb{Q}(\sqrt{2}, \sqrt{3})$ can be put in the form

$$x = a + \frac{1}{2}\alpha(\alpha^2 - 9)b + \frac{1}{2}\alpha(11 - \alpha^2)c + \frac{1}{4}\alpha^2(\alpha^2 - 9)(11 - \alpha^2)d$$

(where $a, b, c, d \in \mathbb{Q}$) in terms of the so-called primitive element α.

Note that α is a root of

$$1 - 10\xi^2 + \xi^4 = 0$$

an equation of degree $m = 4$ in ξ with coefficients in the field $\mathbb{Q}$. (In Chapter 2, we will introduce irreducible polynomials $P_m(\xi)$, of degree m, that generalize the polynomial $1 - 10\xi^2 + \xi^4$.) Consequently, the general element x of $\mathbb{Q}(\sqrt{2}, \sqrt{3})$ can be rewritten as

$$x = a - \frac{5}{2}d + \frac{1}{2}(11c - 9b)\alpha + \frac{1}{2}d\alpha^2 + \frac{1}{2}(b - c)\alpha^3$$

modulo the use of $1 - 10\alpha^2 + \alpha^4 = 0$. Thus, x appears as a polynomial

$$x = x_0 + x_1\alpha + x_2\alpha^2 + x_3\alpha^3$$

of degree $m-1=3$ in α with coefficients

$$x_0 = a - \frac{5}{2}d, \quad x_1 = \frac{1}{2}(11c - 9b), \quad x_2 = \frac{1}{2}d, \quad x_3 = \frac{1}{2}(b-c)$$

in $\mathbb{Q}$. This leads to denote $\mathbb{Q}(\sqrt{2}, \sqrt{3})$ as $\mathbb{Q}[\xi]/\langle 1 - 10\xi^2 + \xi^4 \rangle$ (see Chapter 2).

1.2.5.8. *Example: the fields of real and complex numbers*

The set $\mathbb{R}$ of real numbers and the set $\mathbb{C}$ of complex numbers are commutative fields, with an uncountable infinite number of elements, with respect to the addition and multiplication of real and complex numbers, respectively.

The notion of primitive element mentioned in 1.2.5.7 also applies to the field $\mathbb{C}$. The general element z of $\mathbb{C}$ is of the form $z = x + \mathrm{i}y$ (a polynomial of degree $m-1=1$ in the primitive element $\alpha = \mathrm{i}$ with coefficients x and y in the field $\mathbb{R}$) where α is a root of $1+\xi^2=0$, an equation of degree $m=2$ in ξ with coefficients in $\mathbb{R}$. According to the notion of extension of field introduced in Chapter 2, we can denote $\mathbb{C}$ as $\mathbb{R}[\xi]/\langle 1+\xi^2 \rangle$ (see 2.3.8.2).

1.2.5.9. *Example: the field of quaternions*

The set $\mathbb{H}$ of Hamilton quaternions endowed with the addition $+$ and multiplication $\times$ of quaternions is a field.

For the sake of clarity, some details are in order. The set $\mathbb{H}$ consists of the elements

$$q = t + x\mathrm{i} + y\mathrm{j} + z\mathrm{k}$$

where t, x, y, z belong to $\mathbb{R}$ and $\mathrm{i}, \mathrm{j}, \mathrm{k}$ are such that

$$\mathrm{i} \times \mathrm{i} = \mathrm{j} \times \mathrm{j} = \mathrm{k} \times \mathrm{k} = -1$$

$$\mathrm{i} \times \mathrm{j} = -\mathrm{j} \times \mathrm{i} = \mathrm{k}, \quad \mathrm{j} \times \mathrm{k} = -\mathrm{k} \times \mathrm{j} = \mathrm{i}, \quad \mathrm{k} \times \mathrm{i} = -\mathrm{i} \times \mathrm{k} = \mathrm{j}$$

The addition $q_1 + q_2$ and the multiplication $q_1 \times q_2$ of two quaternions q_1 and q_2 are defined in a way similar to the one for complex numbers by taking into account the relations on i, j, k just mentioned. With evident notations, we obtain

$$q_1 + q_2 = t_1 + t_2 + (x_1 + x_2)\mathrm{i} + (y_1 + y_2)\mathrm{j} + (z_1 + z_2)\mathrm{k}$$

for the addition and

$$\begin{aligned} q_1 \times q_2 = t_1t_2 - x_1x_2 - y_1y_2 - z_1z_2 \\ + (x_1t_2 + t_1x_2 + y_1z_2 - z_1y_2)\mathrm{i} \\ + (y_1t_2 + t_1y_2 + z_1x_2 - x_1z_2)\mathrm{j} \\ + (z_1t_2 + t_1z_2 + x_1y_2 - y_1x_2)\mathrm{k} \end{aligned}$$

for the multiplication. Clearly, $(\mathbb{H}, +)$ is an Abelian group with 0 (corresponding to $t = x = y = z = 0$) as zero element and $(\mathbb{H}^*, \times)$ is a non-commutative group with 1 (corresponding to $t - 1 = x = y = z = 0$) as unity element. Furthermore, the distributivity of the multiplication with respect to the addition can be easily checked. Therefore, $(\mathbb{H}, +, \times)$ is a field. This infinite field, with an uncountable infinite number of elements, is not commutative.

From a practical point of view, it is useful to substitute matrix calculus to quaternion calculus. In this respect, let

$$M(q) = \begin{pmatrix} t + \mathrm{i}x & -y - \mathrm{i}z \\ y - \mathrm{i}z & t - \mathrm{i}x \end{pmatrix}$$

where i is the pure imaginary. It is easy to show that

$$M(q_1 + q_2) = M(q_1) + M(q_2), \quad M(q_1 \times q_2) = M(q_1)M(q_2)$$

Thus, the quaternion field $\mathbb{H}$ is isomorphic to the matrix field $\{M(q) \mid q \in \mathbb{H}\}$ (the $+$ and $\times$ laws of $\mathbb{H}$ correspond to the matrix addition and the matrix multiplication, respectively).

1.2.5.10. *Remark*

The notion of *primitive element* of a finite or infinite field is evoked on several occasions in the preceding examples. A precise definition of a primitive element is given in section 2.3.5 of Chapter 2 for a finite field. At this stage, it is sufficient to give the following definition. Loosely speaking, a primitive element α of a finite field is an element such that every element of the field can be expressed as a power of α (see the example 1.2.5.3).

1.2.6. *Sub-field of a field*

1.2.6.1. *Sub-field*

DEFINITION 1.19.– Let $(\mathbb{K}, +, \times)$ be a field and let $\mathbb{J}$ be a subset of the set $\mathbb{K}$ containing the elements 0 and 1 of $\mathbb{K}$. If the set $\mathbb{J}$ endowed with the addition $+$ and the multiplication $\times$ laws of $(\mathbb{K}, +, \times)$ is a field, then $(\mathbb{J}, +, \times)$ is said to be a *sub-field* of $(\mathbb{K}, +, \times)$. We adopt the notation $\mathbb{J} \subset \mathbb{K}$ to indicate that $(\mathbb{J}, +, \times)$ is a sub-field of $(\mathbb{K}, +, \times)$.

The field $(\mathbb{K}, +, \times)$ is a trivial or improper sub-field of $(\mathbb{K}, +, \times)$. The other sub-fields of $(\mathbb{K}, +, \times)$, if any, are called non-trivial or proper sub-fields of $(\mathbb{K}, +, \times)$. A field that has no proper sub-fields is called a *prime field* (see also 2.1.2.4).

1.2.6.2. *Example:* $\mathbb{Z}_p$, *with* p *prime*

The sole sub-field of the field $\mathbb{Z}_p$ (p prime) is $\mathbb{Z}_p$ (also denoted as $\mathbb{F}_p$). The field $\mathbb{Z}_p$ does not have any proper (i.e. different from $\mathbb{Z}_p$) sub-field. It is a prime field.

1.2.6.3. *Example:* $\mathbb{Q} \subset \mathbb{Q}(\sqrt{2})$

The field $\mathbb{Q}$ of rational numbers is a proper sub-field of $\mathbb{Q}(\sqrt{2})$. Note that the field $\mathbb{Q}$ has no proper sub-fields. It is a prime field.

1.2.6.4. *Example:* $\mathbb{Q}(\sqrt{2}) \subset \mathbb{R}$

The field $\mathbb{Q}(\sqrt{2})$ is the smallest proper sub-field of $\mathbb{R}$ containing $\sqrt{2}$.

1.2.6.5. *Example:* $\mathbb{R} \subset \mathbb{C}$

The field $\mathbb{R}$, the field of real numbers, is a proper sub-field of $\mathbb{C}$, the field of complex numbers.

1.2.6.6. *Example:* $\mathbb{C} \subset \mathbb{H}$

The field $\mathbb{C}$ of complex numbers is a proper sub-field of the field $\mathbb{H}$ of usual quaternions. It is clear that $\mathbb{Q} \subset \mathbb{Q}(\sqrt{2}) \subset \mathbb{R} \subset \mathbb{C} \subset \mathbb{H}$.

1.2.6.7. *Example:* $\overline{\mathbb{Q}} \subset \mathbb{C}$

By definition, an algebraic number is a solution of a polynomial equation of type

$$a_0 + a_1 x + \cdots + a_{n-1} x^{n-1} + x^n = 0, \quad a_0, a_1, \cdots, a_{n-1} \in \mathbb{Q}$$

where $n \in \mathbb{N}_1$. The set of algebraic numbers constitutes a field, noted $\overline{\mathbb{Q}}$, with respect to the addition and multiplication of complex numbers. The field $\overline{\mathbb{Q}}$ is a proper sub-field of $\mathbb{C}$.

1.2.7. *Characteristic of a field*

The definition of the characteristic of a unitary ring applies to a field since a field is a particular unitary ring. In terms of field, we have the following formulation.

1.2.7.1. *Characteristic*

DEFINITION 1.20.– The *characteristic* of a field $(\mathbb{K}, +, \times)$ is the smallest positive integer p ($p \geq 2$) such that

$$\forall x \in \mathbb{K} : p \times x = 0 \Leftrightarrow 1 + 1 + \cdots + 1 = 0$$

where the sum contains p terms. If $1 + 1 + \cdots + 1 \neq 0$ whatever the number of 1 in the sum is, then the field is said to be of characteristic 0.

1.2.7.2. *Example:* $\mathbb{Z}_p$*, with* p *prime*

The field $\mathbb{Z}_p$ (p prime) is a field of characteristic p. We will see that there are other fields of characteristic p.

1.2.7.3. *Example:* $\mathbb{Q}$, $\mathbb{R}$, $\mathbb{C}$ *and* $\mathbb{H}$

The field of rational numbers $\mathbb{Q}$, the field of real numbers $\mathbb{R}$, the field of complex numbers $\mathbb{C}$ and the field of quaternions $\mathbb{H}$ are fields of characteristic 0.

1.2.7.4. *Possible values of the characteristic of a field*

PROPOSITION 1.8.– The characteristic of a field is either zero (for infinite fields) or a prime number (for finite fields).

In other words, if a field has a non-vanishing characteristic, then its characteristic is a prime number and the field is finite. A field of characteristic 2 is called a binary field (the field $\mathbb{F}_2$ is the smallest of the binary fields).

1.2.7.5. *Characteristic of two isomorphic fields*

PROPOSITION 1.9.– Two isomorphic fields have the same characteristic.

1.2.7.6. *Characteristic of a sub-field*

PROPOSITION 1.10.– Let $\mathbb{J}$ be a proper sub-field of a field $\mathbb{K}$. The fields $\mathbb{J}$ and $\mathbb{K}$ have the same characteristic.

2

Galois Fields

This chapter is devoted to finite fields (fields with a finite number of elements greater than or equal to 2) also called Galois fields. A finite field is a particular finite ring. Indeed, a finite unitary commutative ring whose non-zero elements form a multiplicative group is a finite field.

Finite fields play an important role in discrete mathematics (number theory, combinatorics, group theory, Galois theory, algebraic and finite geometries) as well as in classical and quantum information (tomography, cryptography, coding theory, error correction, quantum computing).

Relevant references are listed in the sections *Mathematical literature: rings and fields* and *Useful web links* of the bibliography. The sections *Mathematical literature: number theory* and *Theoretical physics literature: MUBs* contain useful references concerning Gaussian sums over Galois fields. If necessary, as a prerequisite for this chapter, the reader can consult Chapter 5 where some basic elements on number theory and group theory are given.

2.1. Generalities

2.1.1. *Wedderburn's theorem*

PROPOSITION 2.1.– A finite field is necessarily commutative.

PROOF.– The proof of this important theorem can be found in textbooks of pure mathematics. □

Therefore, all finite fields are commutative. There is no non-commutative finite field. Thus, as far as finite fields are concerned, there is no difference between the definitions involving or not the axiom of commutativity for the multiplication law.

2.1.2. *Galois field*

2.1.2.1. *Finite field*

DEFINITION 2.1.– A finite field is called a *Galois field.*

In more detail, a set $\mathbb{K}$ with a finite number of elements, endowed with two internal composition laws, denoted as $+$ with 0 as the neutral element and $\times$ with 1 as the neutral element, such that

– $\mathbb{K}$ is a commutative (or Abelian) group for the law $+$

– $\mathbb{K}^* = \mathbb{K} \setminus \{0\}$ is a group for the law $\times$

– the law $\times$ is distributive with respect to the law $+$

is a Galois field (see section 1.2.1.1 of Chapter 1 for the corresponding axioms). According to Wedderburn's theorem, a Galois field is necessarily a commutative field.

The naming "Galois field" is largely used (in honor of the French mathematician Évariste Galois). Several notations exist for a Galois field. A Galois field of cardinal q is generally denoted as $\mathbb{GF}(q)$, where $\mathbb{GF}$ stands for Galois field, or $\mathbb{F}_q$. In the present book, we use both notations.

2.1.2.2. *Cardinal of a Galois field*

PROPOSITION 2.2.– The cardinal q of a Galois field is necessarily of the form $q = p^m$ where p is a prime number ($p = 2, 3, 5, \cdots$) and m a positive integer ($m \in \mathbb{N}_1$). Furthermore, for every prime number p and integer m in $\mathbb{N}_1$, there exists a Galois field containing p^m elements.

Consequently, there exists no Galois field with a number of elements different from a prime power. A field with p^m elements where $m \geq 2$ will be denoted as $\mathbb{GF}(p^m)$. In the case $m = 1$, a field with p elements will be denoted as $\mathbb{F}_p$.

2.1.2.3. *Galois fields of the same cardinal*

PROPOSITION 2.3.– All Galois fields of the same cardinal are isomorphic. Thus, for any prime number p (p even or odd) and any integer m greater than or equal to 1, there exists one Galois field (and only one, up to an isomorphism) of cardinal p^m.

A Galois field is thus entirely determined by its cardinal. Therefore, all Galois fields with the same cardinal p^m (p prime, $m \geq 1$) are denoted by the same symbol, namely, either $\mathbb{GF}(p^m)$ for $m \geq 2$ or $\mathbb{F}_p$ for $m = 1$.

As an immediate result of this proposition, we have the following corollary related to the case $m = 1$.

2.1.2.4. *Prime field*

PROPOSITION 2.4.– Any Galois field $\mathbb{F}_p$ of cardinal p with p a prime number is isomorphic to $\mathbb{Z}/p\mathbb{Z}$.

Therefore, a Galois field of cardinal p with p a prime number can be denoted as $\mathbb{F}_p$ or $\mathbb{Z}/p\mathbb{Z}$ or simply $\mathbb{Z}_p$. Such a field is referred to as a *prime field* (it has no proper sub-fields). All other Galois fields are of cardinal p^m where p is a prime number and m an integer greater than or equal to 2; they are denoted as $\mathbb{GF}(p^m)$ or $\mathbb{F}_{p^m}$ in the literature. As already mentioned, here we will use the notations $\mathbb{F}_p$ (for $m = 1$) and $\mathbb{GF}(p^m)$ (for $m \geq 2$).

2.1.2.5. *Example:* $\mathbb{F}_2$, $\mathbb{F}_3$ *and* $\mathbb{F}_5$

The two fields of lowest cardinal are $\mathbb{F}_2$ and $\mathbb{F}_3$. The addition and multiplication tables are given in Tables 2.1 and 2.2 for the field $\mathbb{F}_2$ (or $\mathbb{Z}_2$) of cardinal 2 and in Tables 2.3 and 2.4 for the field $\mathbb{F}_3$ (or $\mathbb{Z}_3$) of cardinal 3. The field $\mathbb{F}_2$ is the smallest Galois field and $\mathbb{F}_3$ is the smallest Galois field with an odd number of elements. The next field, following $\mathbb{F}_2$ and $\mathbb{F}_3$, of prime cardinal is $\mathbb{F}_5$ (or $\mathbb{Z}_5$), for which the addition and multiplication tables are given in Tables 1.13 and 1.14, respectively.

2.1.2.6. *Example: a field of cardinal* 5^2

We continue with a field of prime power cardinal. It is easy to check that the set $\{x = a + b\sqrt{2} \mid a \in \mathbb{F}_5,\ b \in \mathbb{F}_5\}$ is a field under the addition and multiplication modulo 5. The inverse for the addition law of an arbitrary element $a + b\sqrt{2}$ is $(-a) + (-b)\sqrt{2}$ where $-a$ and $-b$ are the additive inverses in $\mathbb{F}_5$ of a and b, respectively, and the inverse for the multiplication law of a non-zero element $a + b\sqrt{2}$ is $a' + b'\sqrt{2}$ where

$$a' = \frac{a}{a^2 - 2b^2}, \quad b' = -\frac{b}{a^2 - 2b^2}$$

The addition and multiplication tables of this field of cardinal $25 = 5^2$ can be easily set up. This field is isomorphic to all the fields of cardinal 25.

Note that the general element x of the field under consideration is of the form $x = a + b\alpha$ where $a, b \in \mathbb{F}_5$ and $\alpha = \sqrt{2}$ is a root of $-2 + \xi^2 = 0$ (or $3 + \xi^2 \equiv 0 \bmod 5$), an equation of degree $m = 2$ in ξ with coefficients in the field $\mathbb{F}_5$. The significance of this kind of remark here and elsewhere will become clear in sections 2.2 and 2.3 when we deal with the notions of primitive element and irreducible polynomial over a field $\mathbb{F}_p$. With regard to this, the field under consideration can be denoted as $\mathbb{F}_5[\xi]/\langle 3 + \xi^2 \rangle$ (see 2.3.8.10).

2.1.2.7. *Characteristic of* $\mathbb{GF}(p^m)$

PROPOSITION 2.5.– The characteristic of the Galois field $\mathbb{GF}(p^m)$, with p prime and m positive integer ($m \in \mathbb{N}_1$), is equal to p.

Note that

$$\text{charact}(\mathbb{GF}(p^m)) = \text{charact}(\mathbb{F}_p) = p$$

The number p^m of elements (sometimes referred to as the order) of a Galois field is a power of its characteristic p. For $m = 1$, the number of elements and the characteristic of $\mathbb{F}_p$ are equal.

2.1.2.8. *An identity in* $\mathbb{GF}(p^m)$

PROPOSITION 2.6.– In a Galois field $\mathbb{GF}(p^m)$ of characteristic p, with p prime and m positive integer ($m \in \mathbb{N}_1$), we have the remarkable identity

$$\forall x \in \mathbb{GF}(p^m),\ \forall y \in \mathbb{GF}(p^m) : (x \pm y)^p \equiv x^p \pm y^p \bmod p$$

and more generally,

$$\forall x \in \mathbb{GF}(p^m),\ \forall y \in \mathbb{GF}(p^m) : (x \pm y)^{p^n} \equiv x^{p^n} \pm y^{p^n} \bmod p$$

where n is a non-negative integer (x^s is defined as $x \times x \times \cdots \times x$ with $s \geq 1$ factors).

PROOF.– The proof of the first part of the proposition is similar to the one given in 1.1.10.4 for a ring of characteristic p with p prime (observe that for $p = 2$, $(x \pm y)^2 \equiv x^2 + y^2 \equiv x^2 - y^2 \bmod 2$). Repeated application of this result leads to the second part of the proposition. □

2.2. Extension of a field: a typical example

For $m \geq 2$, the field $\mathbb{GF}(p^m)$ is not of the $\mathbb{Z}_d$ type for $d = p^m$, but can be built from the field $\mathbb{F}_p$ isomorphic to $\mathbb{Z}_p$. In section 2.3, we will show how to construct the field $\mathbb{GF}(p^m)$, with p a prime number and m an integer greater than or equal to 2, from the knowledge of the field $\mathbb{F}_p$. This construction corresponds to a *Galois extension* of the field $\mathbb{F}_p$ by an element that does not belong to $\mathbb{F}_p$.

Rather than starting with the general case, we will begin with an example. In the present section, we will construct the field $\mathbb{GF}(2^2)$ from the field $\mathbb{F}_2$. Here, $p = 2$ and $m = 2$ ($\Rightarrow p^m = 4$). The field $\mathbb{GF}(2^2)$, which is not isomorphic to $\mathbb{Z}_4$, can be built from the field $\mathbb{F}_2$, which is isomorphic to $\mathbb{Z}_2$ (see Tables 2.1 and 2.2 for the addition and multiplication tables of $\mathbb{F}_2$, respectively). The construction of $\mathbb{GF}(2^2)$ can be achieved in three steps.

+	0	1
0	0	1
1	1	0

Table 2.1. *Addition table for the field $\mathbb{F}_2$, noted $\mathbb{Z}_2$ too*

×	0	1
0	0	0
1	0	1

Table 2.2. *Multiplication table for the field $\mathbb{F}_2$, noted $\mathbb{Z}_2$ too*

Step 1. The first step is to search for a polynomial $P_2(\xi)$ of degree $m = 2$

$$P_2(\xi) = a + b\xi + \xi^2, \quad a \in \mathbb{F}_2, \quad b \in \mathbb{F}_2$$

that cannot be factored into the product of two polynomials of degree 1 with coefficients in $\mathbb{F}_2$. Such a polynomial, in the polynomial ring $\mathbb{F}_2[\xi]$, is said to be a *monic irreducible polynomial* or *prime polynomial*, a notion to be developed in the following; at this level, it is sufficient to say that *monic* signifies that the coefficient of the term with the highest degree in $P_2(\xi)$ is equal to 1 and *irreducible* means that $P_2(\xi)$ cannot be factored as $(\xi - \xi_0)(\xi - \xi_1)$ where ξ_0 and ξ_1 belong to $\mathbb{F}_2$. The various candidates for $P_2(\xi)$ are

$$\xi^2 = \xi\xi,\ \xi + \xi^2 = \xi(1+\xi),\ 1+\xi^2 \equiv (1+\xi)(1+\xi) \bmod 2,\ 1+\xi+\xi^2$$

+	0	1	2
0	0	1	2
1	1	2	0
2	2	0	1

Table 2.3. *Addition table for the field $\mathbb{F}_3$, noted $\mathbb{Z}_3$ too*

×	0	1	2
0	0	0	0
1	0	1	2
2	0	2	1

Table 2.4. *Multiplication table for the field $\mathbb{F}_3$, noted $\mathbb{Z}_3$ too*

$P_2(\xi) = a + b\xi + \xi^2$	$P_2(\xi)$	acceptable
$0 + 0 + \xi^2$	ξ^2	no
$0 + \xi + \xi^2$	$\xi(1+\xi)$	no
$1 + 0 + \xi^2$	$1 + \xi^2$	no
$1 + \xi + \xi^2$	$1 + \xi + \xi^2$	yes

Table 2.5. *Possibilities for the polynomial $P_2(\xi)$*

The various possibilities for the coefficients a and b are given in Table 2.5. The sole acceptable possibility is to take

$$P_2(\xi) = 1 + \xi + \xi^2$$

because the latter polynomial cannot be factored into the product of two polynomials of degree 1 with coefficients in $\mathbb{F}_2$. In the present case, an equivalent way to obtain $1 + \xi + \xi^2$ is to look for a polynomial

$P_2(\xi) = a + b\xi + \xi^2$ with a and b in $\mathbb{F}_2$ such that the equation $P_2(\xi) = 0$ has no solution in $\mathbb{F}_2$ ($\Leftrightarrow$ $P_2(\xi)$ cannot be factored over $\mathbb{F}_2$); only the polynomial $1 + \xi + \xi^2$ has no roots in $\mathbb{F}_2$ ($\xi = 0$ and $\xi = 1$ are not solutions of $1 + \xi + \xi^2 = 0$) while the other polynomials in Table 2.5 admit solutions in $\mathbb{F}_2$.

Step 2. The next step amounts to introduce a solution α of $P_2(\xi) = 0$ (solution modulo 2). By construction, this solution (satisfying $1+\alpha+\alpha^2 = 0$) does not belong to $\mathbb{F}_2$. At this point, it is important to note that it is unnecessary to exhibit the detailed expression of α which is $\alpha = \alpha_+$ or $\alpha = \alpha_-$ with

$$\alpha_\pm = \frac{1}{2}(-1 \pm \sqrt{-3})$$

The root α, called primitive element (a notion to be precised later), constitutes one of the elements of $\mathbb{GF}(2^2)$. All the elements x of $\mathbb{GF}(2^2)$ are then given in terms of the primitive element α by

$$x = x_0 + x_1\alpha, \quad x_0 \in \mathbb{F}_2, \quad x_1 \in \mathbb{F}_2$$

that is to say by a polynomial in α of degree $m-1 = 1$. Thus, the four elements (or residue classes) x of the field $\mathbb{GF}(2^2)$ are

$$x = 0,\ 1,\ \alpha,\ 1 + \alpha$$

Note that the element $1 + \alpha$ of $\mathbb{GF}(2^2)$ can be written as

$$1 + \alpha \equiv \alpha^2 \bmod 2 \ \Leftrightarrow\ 1 + \alpha + \alpha^2 = 0$$

Furthermore, we have $\alpha^3 \equiv 1 \bmod 2$ since

$$\alpha^2 \equiv 1 + \alpha \bmod 2 \ \Leftrightarrow\ \alpha^3 \equiv \alpha + \alpha^2 \equiv 1 \bmod 2$$

Therefore, the elements x of $\mathbb{GF}(2^2)$ can be taken in the form

$$x = 0,\ 1,\ \alpha,\ 1 + \alpha \quad \text{or} \quad x = 0,\ 1,\ \alpha,\ \alpha^2 \quad \text{or} \quad x = 0,\ \alpha,\ \alpha^2,\ \alpha^3$$

which can be combined through the laws $+$ and $\times$ modulo 2 together with the constraint $1 + \alpha + \alpha^2 = 0$.

Step 3. Finally, the addition and multiplication tables of $\mathbb{GF}(2^2)$ are given by Tables 2.6 and 2.7 in terms of 0, 1, α and $1+\alpha$, and by Tables 2.8 and 2.9 in terms of 0, $\alpha^3 = 1$, α and α^2. The following relations illustrate some of the equations used for setting the various tables

$$1+1 = 2 \equiv 0, \quad \alpha + \alpha = 2 \times \alpha \equiv 0$$

$$1 + \alpha + \alpha^2 = 0 \Rightarrow \alpha^2 = -1 - \alpha \equiv -1 - \alpha + 2 + (2 \times \alpha) = 1 + \alpha$$

$$\alpha \times (1+\alpha) = \alpha + \alpha^2 = -1 \equiv -1 + 2 = 1$$

$$\alpha^3 = \alpha \times \alpha^2 \equiv \alpha \times (1+\alpha) \equiv 1, \alpha^4 = \alpha \times \alpha^3 \equiv \alpha \times 1 = \alpha$$

where the additions and multiplications are effected modulo 2 and by taking into account $1 + \alpha + \alpha^2 = 0$.

$+$	0	1	α	$1+\alpha$
0	0	1	α	$1+\alpha$
1	1	0	$1+\alpha$	α
α	α	$1+\alpha$	0	1
$1+\alpha$	$1+\alpha$	α	1	0

Table 2.6. *Addition table for the field $\mathbb{GF}(2^2) = \mathbb{F}_2[\xi]/\langle 1+\xi+\xi^2\rangle$: the elements are taken in the form 0, 1, α and $1+\alpha$ (α is a primitive element); observe that this table coincides with the group table of the group $C_2 \times C_2$*

$\times$	0	1	α	$1+\alpha$
0	0	0	0	0
1	0	1	α	$1+\alpha$
α	0	α	$1+\alpha$	1
$1+\alpha$	0	$1+\alpha$	1	α

Table 2.7. *Multiplication table for the field $\mathbb{GF}(2^2) = \mathbb{F}_2[\xi]/\langle 1+\xi+\xi^2\rangle$: the elements are taken in the form 0, 1, α and $1+\alpha$ (α is a primitive element); observe that the part of the table corresponding to the non-zero elements coincides with the group table of the cyclic group C_3*

The field $\mathbb{GF}(2^2)$ so obtained corresponds to the quotient $\mathbb{F}_2[\xi]/\langle 1+\xi+\xi^2\rangle$ whose elements are the remainder polynomials (or residue classes) arising in the relevant Euclidean divisions. This field is an extension of degree 2 of the field $\mathbb{F}_2$ by the primitive element α, root of the monic irreducible polynomial $P_2(\xi)$. The elements of $\mathbb{GF}(2^2) = \mathbb{F}_2[\xi]/\langle 1+\xi+\xi^2 \rangle$ are combined by effecting the operations $+$ and $\times$ modulo 2 and by taking into account $1 + \alpha + \alpha^2 = 0$.

$+$	0	α^3	α	α^2
0	0	α^3	α	α^2
α^3	α^3	0	α^2	α
α	α	α^2	0	α^3
α^2	α^2	α	α^3	0

Table 2.8. *Addition table for the field $\mathbb{GF}(2^2) = \mathbb{F}_2[\xi]/\langle 1+\xi+\xi^2 \rangle$: the elements are taken in the form 0, α^3, α and α^2 (α is a primitive element); observe that this table coincides with the group table of the group $C_2 \times C_2$*

$\times$	0	α^3	α	α^2
0	0	0	0	0
α^3	0	α^3	α	α^2
α	0	α	α^2	α^3
α^2	0	α^2	α^3	α

Table 2.9. *Multiplication table for the field $\mathbb{GF}(2^2) = \mathbb{F}_2[\xi]/\langle 1+\xi+\xi^2 \rangle$: the elements are taken in the form 0, α^3, α and α^2 (α is a primitive element); observe that the part of the table corresponding to the non-zero elements coincides with the group table of the cyclic group C_3*

To close this example, let us note that the field $\mathbb{F}_2$ is a sub-field of the field $\mathbb{GF}(2^2)$, a result to be generalized according to $\mathbb{F}_p \subset \mathbb{GF}(p^m)$ in the case of an arbitrary field $\mathbb{GF}(p^m)$. Note also that the relation $\alpha^3 = 1$ will be generalized as $\alpha^{p^m-1} = 1$ in the case of $\mathbb{GF}(p^m)$.

We are now in a position to generalize such a construction to the field $\mathbb{GF}(p^m)$.

2.3. Extension of a field: the general case

This section deals with the construction of the Galois field $\mathbb{GF}(p^m)$, with p prime and $m \geq 2$, as a Galois extension of degree m of the Galois field $\mathbb{F}_p$ using an element α, root of a monic irreducible polynomial $P_m(\xi)$ of degree m with coefficients in $\mathbb{F}_p$. We will proceed by introducing in turn every relevant ingredient (irreducible polynomial, reducible polynomial, monic polynomial and prime polynomial).

2.3.1. *Reducible, irreducible and prime polynomials*

Let $\mathbb{K}$ be a commutative field. We denote as $\mathbb{K}[\xi]$ the set of all the polynomials in the indeterminate ξ with coefficients in $\mathbb{K}$. It is clear that $\mathbb{K}[\xi]$ is a commutative ring (but not a field) with respect to the laws $+$ and $\times$ of $\mathbb{K}$. A polynomial $P_m(\xi)$ of degree m ($m \geq 1$) in $\mathbb{K}[\xi]$ that cannot be factored into the product of two polynomials of degree greater than 0 in $\mathbb{K}[\xi]$ is said to be an *irreducible polynomial.* It is a *reducible polynomial* in the opposite case. More precisely, we have the following definition.

2.3.1.1. *Reducible and irreducible polynomials*

DEFINITION 2.2.– Let $\mathbb{K}$ be a commutative field and $P_m(\xi)$ be a polynomial, of degree m ($m \geq 1$), in the commutative ring $\mathbb{K}[\xi]$. One says that

– $P_m(\xi)$ is reducible in $\mathbb{K}[\xi]$ (or reducible over $\mathbb{K}$) if it is the product of two non-constant polynomials of $\mathbb{K}[\xi]$;

– $P_m(\xi)$ is irreducible in $\mathbb{K}[\xi]$ (or irreducible over $\mathbb{K}$) if it is non-constant and cannot be factored as the product of two non-constant polynomials in $\mathbb{K}[\xi]$ (in other words, $P_m(\xi)$ is irreducible if it is non-constant, and in the case where $P_m(\xi)$ is the product of two polynomials in $\mathbb{K}[\xi]$, then one factor of the product is constant).

Note that a monic irreducible polynomial $P_m(\xi)$ over $\mathbb{K}$ does not have non-trivial divisors in $\mathbb{K}[\xi]$ (the sole divisors of $P_m(\xi)$ are 1 and $P_m(\xi)$).

2.3.1.2. *Irreducibility and roots*

PROPOSITION 2.7.– Let $\mathbb{K}$ be a Galois field and $P_m(\xi)$ a polynomial, of degree $m \geq 1$, in $\mathbb{K}[\xi]$. We have

– if $m = 1$, then $P_m(\xi)$ is irreducible in $\mathbb{K}[\xi]$;

– if $m \geq 2$ and if $P_m(\xi)$ is irreducible in $\mathbb{K}[\xi]$, then $P_m(\xi)$ has no roots in $\mathbb{K}$;

– if $m = 2$ or 3 and if $P_m(\xi)$ has no roots in $\mathbb{K}$, then $P_m(\xi)$ is irreducible in $\mathbb{K}[\xi]$.

The reciprocal of the second proposition is false except for $m = 2$ and 3. By combining the second and third propositions, it is clear that a polynomial of degree 2 or 3 in $\mathbb{K}[\xi]$ is irreducible over $\mathbb{K}$ if and only if it has no roots in $\mathbb{K}$. Observe that the product of two irreducible polynomials over $\mathbb{K}$, of degree greater than or equal to 2, is reducible although having no roots in $\mathbb{K}$.

2.3.1.3. *Examples*

– The polynomial $(1+\xi^2)^2$ has no roots in $\mathbb{R}$ and it is reducible in $\mathbb{R}[\xi]$. By contrast, the polynomial $1+\xi^2$ has no roots in $\mathbb{R}$ and it is irreducible in $\mathbb{R}[\xi]$.

– The polynomials $1+\xi+\xi^3$ and $1+\xi^2+\xi^3$ have no roots in $\mathbb{F}_2$ ($\xi = 0$ and $\xi = 1$ are not roots). Therefore, they are irreducible over $\mathbb{F}_2$.

– The polynomial $1+\xi^2+\xi^4$ is reducible over $\mathbb{F}_2$ since

$$1+\xi^2+\xi^4 \equiv (1+\xi+\xi^2)^2 \bmod 2$$

is the product of two polynomials of degree 2 in $\mathbb{F}_2[\xi]$. Observe that $\xi = 0$ and $\xi = 1$ are not roots of $1+\xi^2+\xi^4$ in $\mathbb{F}_2$.

– The polynomial $1+\xi+\xi^3$ is reducible over $\mathbb{F}_3$ since

$$1+\xi+\xi^3 \equiv (2+\xi)(2+\xi+\xi^2) \bmod 3$$

is the product of two polynomials of degrees 1 and 2 in $\mathbb{F}_3[\xi]$. Observe that $\xi = 1$ is a root of $1+\xi+\xi^3$ in $\mathbb{F}_3$.

2.3.1.4. *Monic polynomial and prime polynomial*

DEFINITION 2.3.– Let $\mathbb{K}$ be a field. A *monic polynomial* in $\mathbb{K}[\xi]$ of degree m ($m \in \mathbb{N}_1$) is of the form

$$P_m(\xi) = c_0 + c_1\xi + \cdots + c_{m-1}\xi^{m-1} + \xi^m, \quad c_0, c_1, \cdots, c_{m-1} \in \mathbb{K}$$

(the coefficient of ξ^m is equal to 1). A monic irreducible polynomial is called a *prime polynomial*.

PROPOSITION 2.8.– In the case $\mathbb{K} = \mathbb{F}_p$ (p prime), for any positive integer m, there exists at least one prime polynomial $P_m(\xi)$ of degree m in the ring $\mathbb{F}_p[\xi]$. The number $N(p,m)$ of prime polynomials of degree m in $\mathbb{F}_p[\xi]$ is

$$N(p,m) = \frac{1}{m} \sum_{k,\, k|m} \mu\left(\frac{m}{k}\right) p^k = \frac{1}{m} \sum_{k,\, k|m} \mu(k)\, p^{\frac{m}{k}}$$

where μ is the Möbius function (see Appendix (Chapter 5) for the definition of the Möbius function μ).

2.3.1.5. *Examples*

As trivial examples, the formula for $N(p,m)$ gives

$$N(p,1) = p, \quad N(p,2) = \tfrac{1}{2}p(p-1), \quad N(p,3) = \tfrac{1}{3}p(p-1)(p+1)$$
$$N(p,4) = \tfrac{1}{4}p^2(p-1)(p+1), \quad N(p,5) = \tfrac{1}{5}p(p-1)(p+1)(p^2+1)$$
$$N(p,6) = \tfrac{1}{6}p(p-1)(p+1)(p^3+p-1)$$

The number $N(p,m)$ of prime polynomials $P_m(\xi)$ of degree m in $\mathbb{F}_p[\xi]$ is greater than 1 except for $P_2(\xi)$ in $\mathbb{F}_2[\xi]$ (see some examples in Table 2.10).

p	2	2	2	2	2	2	3	3	3	3	3	3	4	4	4	4
m	1	2	3	4	5	6	1	2	3	4	5	6	1	2	3	4
$N(p,m)$	2	1	2	3	6	9	3	2	8	18	48	116	4	6	20	60

Table 2.10. *Number $N(p,m)$ of prime (i.e. monic + irreducible) polynomials $P_m(\xi)$ of degree m in $\mathbb{F}_p[\xi]$*

We list in Table 2.11 some monic irreducible polynomials (i.e. prime polynomials) $P_m(\xi)$ in $\mathbb{F}_p[\xi]$ useful for the construction of some fields $\mathbb{GF}(p^m)$ in the cases ($p = 2$, $m = 1$ to 6) and ($p = 3$, $m = 1$ to 3).

2.3.1.6. *Factorization of monic polynomials*

PROPOSITION 2.9.– Let $\mathbb{K}$ be a field. Every monic polynomial of $\mathbb{K}[\xi]$, of degree greater than or equal to 1, may be factored into a product of prime polynomials of $\mathbb{K}[\xi]$. The factorization is unique up to the order of the factors.

Prime polynomials in $\mathbb{K}[\xi]$ are analogous to prime integers in $\mathbb{Z}$ (see the fundamental theorem of arithmetic: every integer of $\mathbb{Z}$ may be factored into a product of prime integers). In particular, any monic polynomial in $\mathbb{F}_p[\xi]$ can be factored into a product of powers of prime polynomials of $\mathbb{F}_p[\xi]$; for instance, we have the decompositions

$$1+\xi+\xi^5 \equiv (1+\xi^2+\xi^3)(1+\xi+\xi^2)$$

$$1+\xi+\xi^2+\xi^3+\xi^5+\xi^7 \equiv (1+\xi)^2(1+\xi+\xi^2)(1+\xi^2+\xi^3)$$

$$1+\xi+\xi^2+\xi^3+\xi^4+\xi^8 \equiv (1+\xi)^3(1+\xi+\xi^2)(1+\xi+\xi^3)$$

(mod 2) into prime polynomials of $\mathbb{F}_2[\xi]$.

Monic Irreducible Polynomials $P_m(\xi)$ in $\mathbb{F}_p[\xi]$	p	m
$\xi,\quad 1+\xi$	2	1
$1+\xi+\xi^2$	2	2
$1+\xi+\xi^3,\quad 1+\xi^2+\xi^3$	2	3
$1+\xi+\xi^4,\quad 1+\xi^3+\xi^4,\quad 1+\xi+\xi^2+\xi^3+\xi^4$	2	4
$1+\xi^2+\xi^5,\quad 1+\xi^3+\xi^5$ $1+\xi+\xi^2+\xi^3+\xi^5,\quad 1+\xi+\xi^2+\xi^4+\xi^5$ $1+\xi+\xi^3+\xi^4+\xi^5,\quad 1+\xi^2+\xi^3+\xi^4+\xi^5$	2	5
$1+\xi+\xi^6,\quad 1+\xi^3+\xi^6,\quad 1+\xi^5+\xi^6$ $1+\xi+\xi^2+\xi^4+\xi^6,\quad 1+\xi+\xi^2+\xi^5+\xi^6$ $1+\xi+\xi^3+\xi^4+\xi^6,\quad 1+\xi+\xi^4+\xi^5+\xi^6$ $1+\xi^2+\xi^3+\xi^5+\xi^6,\quad 1+\xi^2+\xi^4+\xi^5+\xi^6$	2	6
$\xi,\quad 1+\xi,\quad 2+\xi$	3	1
$1+\xi^2,\quad 2+\xi+\xi^2,\quad 2+2\xi+\xi^2$	3	2
$1+2\xi+\xi^3,\quad 2+2\xi+\xi^3,\quad 2+\xi+\xi^2+\xi^3$ $1+2\xi^2+\xi^3,\quad 2+\xi^2+\xi^3,\quad 1+\xi+2\xi^2+\xi^3$ $1+2\xi+\xi^2+\xi^3,\quad 2+2\xi+2\xi^2+\xi^3$	3	3

Table 2.11. *Prime (i.e. monic + irreducible) polynomials $P_m(\xi)$ in $\mathbb{F}_p[\xi]$ for the cases ($p=2$, $m=1$ to 6) and ($p=3$, $m=1$ to 3)*

2.3.2. *Examples of (ir)reducible and prime polynomials*

We give examples of reducible, irreducible and prime polynomials for various fields below. Although this chapter is devoted to finite fields, three examples of infinite fields are given in addition to examples of finite fields.

2.3.2.1. *Example: reducible polynomial over $\mathbb{C}$, $\mathbb{R}$ and $\mathbb{Q}$*

The polynomial $-4+\xi^4$ can be factored as

$$-4+\xi^4 = (\mathrm{i}\sqrt{2}+\xi)(-\mathrm{i}\sqrt{2}+\xi)(\sqrt{2}+\xi)(-\sqrt{2}+\xi)$$

Therefore, it is reducible over $\mathbb{C}$. Furthermore,

$$-4 + \xi^4 = (2 + \xi^2)(-2 + \xi^2)$$

shows that $-4 + \xi^4$ is also reducible over $\mathbb{R}$ and $\mathbb{Q}$.

2.3.2.2. *Example: irreducible polynomial over* $\mathbb{Q}$

The polynomial $-2 + \xi^2$ cannot be factored in $\mathbb{Q}[\xi]$: it is irreducible over $\mathbb{Q}$. However, the factorization

$$-2 + \xi^2 = (\sqrt{2} + \xi)(-\sqrt{2} + \xi)$$

in $\mathbb{R}[\xi]$ shows that $-2 + \xi^2$ is reducible over $\mathbb{R}$.

2.3.2.3. *Example: irreducible polynomial over* $\mathbb{Q}$

It can be shown that the polynomial $1 - 10\xi^2 + \xi^4$ cannot be factored in $\mathbb{Q}[\xi]$ (the proof can be achieved by *reductio ad absurdum*). It is thus irreducible over $\mathbb{Q}$.

2.3.2.4. *Example: reducible polynomials over* $\mathbb{F}_2$

The polynomial $1+\xi^2$ in $\mathbb{F}_2[\xi]$ is reducible over $\mathbb{F}_2$. As a matter of fact, we have

$$1 + \xi^2 \equiv 1 + 2\xi + \xi^2 = (1 + \xi)(1 + \xi) \bmod 2$$

which shows that $1+\xi^2$ is the product of two polynomials of degree 1 in $\mathbb{F}_2[\xi]$. Note that $\xi = 1$ is a (double) root of $1 + \xi^2 = 0$ in $\mathbb{F}_2$.

Similarly, the polynomial $1 + \xi^2 + \xi^4 + \xi^6 + \xi^8$ in $\mathbb{F}_2[\xi]$ is reducible over $\mathbb{F}_2$ since

$$1 + \xi^2 + \xi^4 + \xi^6 + \xi^8 \equiv (1 + \xi + \xi^2 + \xi^3 + \xi^4)^2 \bmod 2$$

so that $1 + \xi^2 + \xi^4 + \xi^6 + \xi^8$ is the product of two polynomials of degree 4 in $\mathbb{F}_2[\xi]$. Observe that $1 + \xi^2 + \xi^4 + \xi^6 + \xi^8$ has no roots in $\mathbb{F}_2$. Note that the polynomial $1 + \xi^2 + \xi^4 + \xi^6$, for which $\xi = 1$ is a root (of multiplicity 6) in $\mathbb{F}_2$, is also reducible over $\mathbb{F}_2$.

2.3.2.5. *Example: reducible polynomials over* $\mathbb{F}_3$

First, the polynomial $2 + \xi^2$ in $\mathbb{F}_3[\xi]$ can be factored into two polynomials of degree 1 in $\mathbb{F}_3[\xi]$ according to

$$2 + \xi^2 \equiv 2 + 3\xi + \xi^2 = (1 + \xi)(2 + \xi) \mod 3$$

Therefore, $2 + \xi^2$ is reducible over $\mathbb{F}_3$. Note that $\xi = 1$ and $\xi = 2$ are (simple) roots of $2 + \xi^2 = 0$ in $\mathbb{F}_3$.

Second, the polynomial $2 + \xi^3$ in $\mathbb{F}_3[\xi]$ is reducible over $\mathbb{F}_3$ since

$$2 + \xi^3 \equiv (2 + \xi)^3 \mod 3$$

Note that $\xi = 1$ is a (triple) root of $2 + \xi^3 = 0$ in $\mathbb{F}_3$.

Third, the polynomial $2 + \xi^4$ in $\mathbb{F}_3[\xi]$ is reducible over $\mathbb{F}_3$ since

$$2 + \xi^4 \equiv (1 + \xi)(2 + \xi)(1 + \xi^2) \mod 3$$

Note that $\xi = 1$ and $\xi = 2$ are (simple) roots of $2 + \xi^4 = 0$ in $\mathbb{F}_3$.

2.3.2.6. *Example: the polynomial* $1 + \xi^2$ *in* $\mathbb{F}_p[\xi]$, $p = 2, 3, 5, 7$

The polynomial $1 + \xi^2$ is reducible over $\mathbb{F}_2$ and $\mathbb{F}_5$ since

$$1 + \xi^2 \equiv (1 + \xi)(1 + \xi) \mod 2, \quad 1 + \xi^2 \equiv (2 + \xi)(3 + \xi) \mod 5$$

while it is irreducible over $\mathbb{F}_3$ and $\mathbb{F}_7$ since $1 + \xi^2$ has no roots in $\mathbb{F}_3$ and $\mathbb{F}_7$.

2.3.2.7. *Example: prime polynomials in* $\mathbb{F}_2[\xi]$

The two polynomials

$$\xi, \quad 1 + \xi$$

cannot be factored over $\mathbb{F}_2$ in polynomials of lower degree; they are prime polynomials of degree 1 in $\mathbb{F}_2[\xi]$.

As was already seen, the polynomial

$$1 + \xi + \xi^2$$

is the sole prime polynomial of degree 2 in $\mathbb{F}_2[\xi]$ (this polynomial has no roots in $\mathbb{F}_2$).

The two polynomials

$$1+\xi+\xi^3, \quad 1+\xi^2+\xi^3$$

are prime polynomials of degree 3 in $\mathbb{F}_2[\xi]$. This can be shown in two ways. First, they do not have roots in $\mathbb{F}_2$. Second, the eight monic polynomials of degree 3 over $\mathbb{F}_2$ are

$$\xi^3 = \xi\xi\xi, \quad \xi^2+\xi^3 = \xi\xi(1+\xi), \quad \xi+\xi^3 = \xi(1+\xi^2)$$

$$\xi+\xi^2+\xi^3 = \xi(1+\xi+\xi^2)$$

$$1+\xi^3 \equiv (1+\xi)(1+\xi+\xi^2) \bmod 2$$

$$1+\xi+\xi^2+\xi^3 \equiv (1+\xi)(1+\xi)(1+\xi) \bmod 2$$

$$1+\xi+\xi^3, \quad 1+\xi^2+\xi^3$$

and it is clear that only the last two polynomials cannot be factored in terms of polynomials of lower degree in $\mathbb{F}_2[\xi]$.

The three polynomials

$$1+\xi+\xi^4, \quad 1+\xi^3+\xi^4, \quad 1+\xi+\xi^2+\xi^3+\xi^4$$

are prime polynomials of degree 4 in $\mathbb{F}_2[\xi]$. However, this does not follow from the fact that the latter polynomials have no roots in $\mathbb{F}_2$ (see the polynomial $1+\xi^2+\xi^4$ has no roots in $\mathbb{F}_2$ and is reducible over $\mathbb{F}_2$). In view of 2.3.1.6, this follows from the fact that none of these three polynomials is divisible by ξ, $1+\xi, 1+\xi+\xi^2$ and therefore cannot be factored in $\mathbb{F}_2[\xi]$ (see the polynomial $1+\xi^2+\xi^4$ is divisible by $1+\xi+\xi^2$ and therefore can be factored in $\mathbb{F}_2[\xi]$).

Finally, note that $1+\xi+\xi^m$ $(m \geq 2)$ is a prime polynomial of degree m in $\mathbb{F}_2[\xi]$.

2.3.2.8. *Example: prime polynomials in* $\mathbb{F}_3[\xi]$

The polynomials

$$\xi, \quad 1+\xi, \quad 2+\xi, \quad 1+\xi^2, \quad 2+\xi+\xi^2, \quad 2+2\xi+\xi^m \; (m \geq 2)$$

are prime polynomials in $\mathbb{F}_3[\xi]$.

2.3.2.9. *Example: prime polynomials in* $\mathbb{F}_p[\xi]$

More generally, a prime polynomial in the ring $\mathbb{F}_p[\xi]$ is of the form

$$P_m(\xi) = c_0 + c_1\xi + \cdots + c_{m-1}\xi^{m-1} + \xi^m$$

where $c_0, c_1, \cdots, c_{m-1}$ are in $\mathbb{F}_p$ and such that $P_m(\xi)$ cannot be factored into polynomials over $\mathbb{F}_p$ of degree lower than m. Such a polynomial exists for any positive integer m.

2.3.3. *Quotient field*

2.3.3.1. *Field as a quotient*

PROPOSITION 2.10.– Let $\mathbb{K}$ be a field and $\mathbb{K}[\xi]$ be the ring of polynomials in the indeterminate ξ and with coefficients in $\mathbb{K}$, and let $P_m(\xi)$ be a polynomial of degree m in $\mathbb{K}[\xi]$. Then, the quotient $\mathbb{K}[\xi]/\langle P_m(\xi)\rangle$ is a field if and only if $P_m(\xi)$ is irreducible over $\mathbb{K}$.

Note that the quotient $\mathbb{K}[\xi]/\langle P_m(\xi)\rangle$ is a ring if $P_m(\xi)$ is reducible over $\mathbb{K}$.

2.3.3.2. *Application to* $\mathbb{GF}(p^m)$

The latter general proposition provides a way to construct the Galois field $\mathbb{GF}(p^m)$ with p^m elements from the field $\mathbb{K} = \mathbb{F}_p$ with p elements (p prime number, m positive integer greater than 1). In this context, the prime field $\mathbb{F}_p$ is referred to as a *base field*. Indeed, we have

$$\mathbb{GF}(p^m) = \mathbb{F}_p[\xi]/\langle P_m(\xi)\rangle$$

where $P_m(\xi)$ is a monic irreducible polynomial of degree m in $\mathbb{F}_p[\xi]$. The field $\mathbb{GF}(p^m)$ is the residue class field of the ring of polynomials $\mathbb{F}_p[\xi]$ modulo $\langle P_m(\xi)\rangle$. The p^m elements of $\mathbb{GF}(p^m)$ are thus the residue classes of polynomials with coefficients in $\mathbb{F}_p$ arising from the Euclidean division by $P_m(\xi)$ of polynomials in $\mathbb{F}_p[\xi]$. Therefore, every element x, denoted as $x(\alpha)$ too, of $\mathbb{GF}(p^m) = \mathbb{F}_p[\xi]/\langle P_m(\xi)\rangle$ can be expressed as a polynomial of degree less than m in of the form

$$\begin{aligned} x \text{ or } x(\alpha) &= x_0 + x_1\alpha + \cdots + x_{m-1}\alpha^{m-1} \\ &= \sum_{k=0}^{m-1} x_k\alpha^k, \quad x_0, x_1, \cdots, x_{m-1} \in \mathbb{F}_p \end{aligned}$$

where α is a root of $P_m(\xi)$ (in the sum, the monomials $x_0, x_1\alpha, \cdots$ mean $x_0 \times 1, x_1 \times \alpha, \cdots$, respectively). A useful notation for $x_0+x_1\alpha+\cdots+x_{m-1}\alpha^{m-1}$ is $[x_0\ x_1\ \cdots\ x_{m-1}]$ or $[x_0\ x_1\ \cdots\ x_{m-1}]_\alpha$, a notation that suggests considering x as a vector of components $x_0, x_1, \cdots, x_{m-1}$ in the field $\mathbb{F}_p$. (In this notation, the sum of two elements $x = [x_0\ x_1\ \cdots\ x_{m-1}]$ and $y = [y_0\ y_1\ \cdots\ y_{m-1}]$ reads $x+y = [x_0+y_0\ x_1+y_1\ \cdots\ x_{m-1}+y_{m-1}]$.) Note that the case $m = 1$ corresponds to $\mathbb{GF}(p) = \mathbb{F}_p$.

If $P_m(\xi)$ is reducible over $\mathbb{F}_p$, then $\mathbb{F}_p[\xi]/\langle P_m(\xi)\rangle$ is a ring rather than a field. For instance, $\mathbb{F}_2[\xi]/\langle 1+\xi^2\rangle$ is not a field because $1+\xi^2$ is not an irreducible polynomial over $\mathbb{F}_2$ (since $1+\xi^2 \equiv (1+\xi)^2$ mod 2). It can be checked that $\mathbb{F}_2[\xi]/\langle 1+\xi^2\rangle$ is a ring (the ring $\mathbb{F}_2[\xi]/\langle \xi^2\rangle$, see 1.1.5.4, and the ring $\mathbb{F}_2[\xi]/\langle 1+\xi^2\rangle$ are isomorphic).

The field $\mathbb{GF}(p^m)$ is entirely determined by its cardinal (Card($\mathbb{GF}(p^m)$) $= p^m$): two polynomials of degree m irreducible over $\mathbb{F}_p$ yield two isomorphic fields of cardinal p^m and characteristic p. Moreover, for any p^m (p prime number, m positive integer), there is one and only one (up to isomorphism) Galois field of characteristic p and cardinal p^m.

Any polynomial $P(\xi)$ in $\mathbb{F}_p[\xi]$ can be written as

$$\begin{aligned} P(\xi) = P_m(\xi) &\times (\text{quotient polynomial of } P(\xi) \text{ by } P_m(\xi)) \\ &+ (\text{remainder of the Euclidean division of } P(\xi) \text{ by } P_m(\xi)) \end{aligned}$$

modulo p. The first part of this sum vanishes when $\xi = \alpha$ (root of $P_m(\xi)$) and thus the remainder of the Euclidean division gives an element of $\mathbb{GF}(p^m)$ in terms of α modulo p.

2.3.3.3. *Sum of the elements of* $\mathbb{GF}(p^m)$

PROPOSITION 2.11.– The sum

$$s(p, m) = \sum_{x \in \mathbb{GF}(p^m)} x$$

of all the elements of the Galois field $\mathbb{GF}(p^m)$ is equal to 0, the additive neutral element of $(\mathbb{GF}(p^m), +)$, except for $\mathbb{GF}(2^1) = \mathbb{F}_2$ for which $s(2, 1) = 1$.

PROOF.– The proof follows from repeated application of $s(p, 1) = 0$ for $p \neq 2$ and is trivial for $s(2, 1)$. □

2.3.4. *Group structures*

2.3.4.1. *Additive and multiplicative groups*

PROPOSITION 2.12.– The group structures related to the field $\mathbb{GF}(p^m)$ are of two kinds: a structure of *additive group* for $\mathbb{GF}(p^m)$ with respect to the law $+$ spanned by all the elements of $\mathbb{GF}(p^m)$ and a structure of *multiplicative group* for $\mathbb{GF}(p^m)^* = \mathbb{GF}(p^m) \setminus \{0\}$ with respect to the law $\times$ spanned by all the non-zero elements of $\mathbb{GF}(p^m)$.

– Structure of additive group: we have the group

$$(\mathbb{GF}(p^m), +) \simeq C_p \times C_p \times \cdots \times C_p$$

i.e. $(\mathbb{GF}(p^m), +)$ is isomorphic to the direct product $C_p \times C_p \times \cdots \times C_p$ of order p^m (the direct product group contains m factors C_p, where C_p is the cyclic group of order p).

– Structure of multiplicative group: we have the group

$$(\mathbb{GF}(p^m)^*, \times) \simeq C_{p^m-1}$$

i.e. $(\mathbb{GF}(p^m)^*, \times)$ is isomorphic to the cyclic group C_{p^m-1} of order $p^m - 1$.

PROOF.– For the group $(\mathbb{GF}(p^m), +)$, the proof is based on the facts that any element x of $\mathbb{GF}(p^m)$ can be written as $x = [x_0 \; x_1 \; \cdots \; x_{m-1}]$ and that the group $(\{x_i \in \mathbb{F}_p\}, +)$ is isomorphic to C_p. For the group $(\mathbb{GF}(p^m)^*, \times)$, the proof follows from the writing of any non-zero element of $\mathbb{GF}(p^m)$ as a power of a primitive element. The reader not familiar with groups may consult Chapter 5 for a short introduction to group theory. □

Of course, in the limit case $m = 1$, the preceding results reduce to

$$(\mathbb{F}_p, +) \simeq C_p$$

i.e. $(\mathbb{F}_p, +)$ is isomorphic to the cyclic group C_p of order p, and

$$({\mathbb{F}_p}^*, \times) \simeq C_{p-1}$$

i.e. $({\mathbb{F}_p}^*, \times)$ is isomorphic to the cyclic group C_{p-1} of order $p - 1$ itself isomorphic to $(\mathbb{Z}_{p-1}, +)$.

Note that the elements of the multiplicative group $(\mathbb{GF}(p^m)^*, \times)$ are made up of all the units of $\mathbb{GF}(p^m)$.

2.3.4.2. *Example: the field* $\mathbb{GF}(2^2)$

It can be easily checked that $(\mathbb{GF}(2^2), +)$ is an Abelian group isomorphic to the Klein four-group V, itself isomorphic to the direct product $C_2 \times C_2$ (compare Table 2.6 or 2.8 with Table 5.8). Furthermore, from Table 2.9, we can extract Table 2.12 which corresponds to the group $(\mathbb{GF}(2^2)^*, \times)$. This group is an Abelian group isomorphic to the cyclic group C_3. It is generated by the element α (the group consists of the elements $\alpha, \alpha^2, \alpha^3 = 1$), a fact that justifies the naming primitive element for α, a notion to be further precised in section 2.3.5.1. To summarize, we have

$$(\mathbb{GF}(2^2), +) \simeq C_2 \times C_2, \quad (\mathbb{GF}(2^2)^*, \times) \simeq C_3$$

in terms of cyclic groups.

$(\mathbb{GF}(2^2)^*, \times)$	α	α^2	α^3
α	α^2	α^3	α
α^2	α^3	α	α^2
α^3	α	α^2	α^3

Table 2.12. *Group table of* $(\mathbb{GF}(2^2)^*, \times)$*: the elements of the group* $(\mathbb{GF}(2^2)^*, \times)$ *are taken in the form* α, α^2 *and* α^3 ($\alpha^2 = 1 + \alpha$ *and* $\alpha^3 = 1$ *where* α *is a primitive element)*

2.3.4.3. *Example: the field* $\mathbb{F}_5$

Table 1.13 shows that $(\mathbb{F}_5, +)$ is isomorphic to the cyclic group C_5: the isomorphism $(\mathbb{F}_5, +) \simeq C_5$ is described by

$$1 \to a, \quad 2 \to a^2, \quad 3 \to a^3, \quad 4 \to a^4, \quad 0 \to a^5$$

where a is a generator of the group C_5. From Table 1.14, we can extract Table 2.13 which is the group table of the cyclic group C_4. To summarize, we have

$$(\mathbb{F}_5, +) \simeq C_5, \quad ({\mathbb{F}_5}^*, \times) \simeq C_4$$

in terms of cyclic groups.

$(\mathbb{F}_5{}^*, \times)$	$2^1 = 2$	$2^2 = 4$	$2^3 = 3$	$2^4 = 1$
$2^1 = 2$	4	3	1	2
$2^2 = 4$	3	1	2	4
$2^3 = 3$	1	2	4	3
$2^4 = 1$	2	4	3	1

Table 2.13. *Group table of* $(\mathbb{F}_5{}^*, \times)$*: the elements of the group* $(\mathbb{F}_5{}^*, \times)$ *are taken in the form* $2^1 = 2$, $2^2 = 4$, $2^3 = 3$ *and* $2^4 = 1$ *(2 is a primitive element)*

2.3.4.4. *Remark*

It is to be emphasized that the structure of the additive group $(\mathbb{GF}(p^m), +)$ does not depend on the monic irreducible polynomial $P_m(\xi)$ used to construct the field $\mathbb{GF}(p^m) = \mathbb{F}_p[\xi]/\langle P_m(\xi)\rangle$. This can be shown as follows. Let $x(\alpha) = x_0 + x_1\alpha + \cdots + x_{m-1}\alpha^{m-1}$ be an element of $\mathbb{GF}(p^m)$, let us consider the set

$$S_{p^m} = \{[x_0\ x_1\ \cdots\ x_{m-1}] \mid x_i \in \mathbb{F}_p,\ i = 0, 1, \cdots, m-1\}$$

and let us endow this set with the law $\oplus$ defined as

$$\begin{aligned}&[x_0\ x_1\ \cdots\ x_{m-1}] \oplus [x_0'\ x_1'\ \cdots\ x_{m-1}'] \\ &\quad = [x_0 + x_0'\ x_1 + x_1'\ \cdots\ x_{m-1} + x_{m-1}']\end{aligned}$$

where the law $+$ refers to the additive law of the field $\mathbb{F}_p$. The set S_{p^m}, of cardinal p^m, is a group with respect to the law $\oplus$. The structure of this group obviously does not depend on the polynomial $P_m(\xi)$. It is clear that this group turns out to be isomorphic to $(\mathbb{GF}(p^m), +)$.

2.3.5. *Primitive element and primitive polynomial*

2.3.5.1. *Primitive element*

DEFINITION 2.4.– The element x of the Galois field $\mathbb{GF}(p^m)$ is called a *primitive element* if it is a generator of the cyclic group $(\mathbb{GF}(p^m)^*, \times)$. In other words,

$$\nexists j < p^m - 1 \mid x^j = 1$$

and all the powers x^i for $i = 1, 2, \cdots, p^m - 1$ are distinct.

2.3.5.2. *Primitive element, sum and product*

If α is a primitive element of $\mathbb{GF}(p^m)$, any non-zero element of $\mathbb{GF}(p^m)$ can be expressed as a power of α (α necessarily belongs to $\mathbb{F}_p$ for $m = 1$ and to $\mathbb{GF}(p^m) \setminus \mathbb{F}_p$ for $m \geq 2$). Then, the p^m elements of the field $\mathbb{GF}(p^m)$ are

$$0,\ \alpha^1,\ \alpha^2,\ \cdots,\ \alpha^{p^m-2},\quad \alpha^{p^m-1} = 1$$

(α is of order $p^m - 1$). These elements are the distinct roots of $\xi^{p^m} - \xi = 0$ (see section 2.5.2.2). The $p^m - 1$ elements of the group $(\mathbb{GF}(p^m)^*, \times)$ are then

$$\alpha^1,\ \alpha^2,\ \cdots,\ \alpha^{p^m-2},\ \alpha^{p^m-1} = 1$$

(the cyclic group $(\mathbb{GF}(p^m)^*, \times)$ generated by α is often denoted as $\langle\alpha\rangle$ in the literature).

The sum (addition) and product (multiplication) of two arbitrary elements of $\mathbb{GF}(p^m)$ are either 0 or a power of α. This can be exemplified with the field $\mathbb{GF}(2^2) = \mathbb{F}_2[\xi]/\langle 1 + \xi + \xi^2\rangle$. Let α be a root of the primitive polynomial $1 + \xi + \xi^2$. Then, the addition and multiplication tables of $\mathbb{GF}(2^2)$ in terms of 0 and powers of α are given by Tables 2.8 and 2.9, respectively.

2.3.5.3. *Remarks*

– Note that if α is a primitive element of $\mathbb{GF}(p^m)$, then its inverse α^{-1} is a primitive element too. If $m \geq 2$, a primitive element of $\mathbb{GF}(p^m) = \mathbb{F}_p[\xi]/\langle P_m(\xi)\rangle$ is not necessarily a root of the prime polynomial $P_m(\xi)$; if a primitive element is a root of $P_m(\xi)$, the prime polynomial $P_m(\xi)$ is called a *primitive polynomial* (see section 2.3.5.8).

– In every Galois field $\mathbb{GF}(p^m)$, there exists at least one element of order $p^m - 1$.

– In a field $\mathbb{GF}(p^m)$, there are either no elements of order n ($1 < n < p^m - 1$) or $\varphi(n)$ elements of order n in $\mathbb{GF}(p^m)^*$ (see Appendix (Chapter 5) for the definition of the Euler function φ). More precisely, if n ($1 < n < p^m - 1$) does not divide $p^m - 1$, then there are no elements of order n (if $p^m - 1$ is a prime number, then all the elements of $\mathbb{GF}(p^m)$ except 0 and 1 are primitive elements). Conversely, if $p^m - 1$ is divisible by n ($1 < n < p^m - 1$), then there are $\varphi(n)$ elements of order n in $\mathbb{GF}(p^m)^*$ which are not primitive elements of $\mathbb{GF}(p^m)$.

– A primitive element of $\mathbb{GF}(p^m)$ cannot belong to a sub-field of $\mathbb{GF}(p^m)$.

2.3.5.4. *Counter-examples*

– Let $x = 2$ be an element of the field $\mathbb{GF}(7^1) = \mathbb{F}_7$. This element is not a primitive element of $\mathbb{F}_7$ since $x^3 \equiv 1 \bmod 7$.

– Let $x = \alpha$ be a root of the prime polynomial $1 + \xi^2$ over $\mathbb{F}_3[\xi]$. The element x is not a primitive element of $\mathbb{GF}(3^2) = \mathbb{F}_3[\xi]/\langle 1 + \xi^2 \rangle$ since $x^4 \equiv 1 \bmod 3$.

– Let $x = \alpha$ be a root of the prime polynomial $1 + \xi + \xi^2 + \xi^3 + \xi^4$ over $\mathbb{F}_2$. The element x is not a primitive element of $\mathbb{GF}(2^4) = \mathbb{F}_2[\xi]/\langle 1+\xi+\xi^2+\xi^3+\xi^4 \rangle$ since $x^5 = 1$.

– Let $x = \alpha$ be a root of the prime polynomial $2 + 2\xi + \xi^3$ over $\mathbb{F}_3$. The element x is not a primitive element of $\mathbb{GF}(3^3) = \mathbb{F}_3[\xi]/\langle 2 + 2\xi + \xi^3 \rangle$ since $x^{13} \equiv 1 \bmod 3$.

– Let $x = \alpha^2$ where α is a root of the prime polynomial $1 + 2\xi + \xi^3$ over $\mathbb{F}_3$. The element x is not a primitive element of $\mathbb{GF}(3^3) = \mathbb{F}_3[\xi]/\langle 1+2\xi+\xi^3 \rangle$ since $x^{13} \equiv 1 \bmod 3$.

In the five preceding examples, the element x does not generate the corresponding cyclic group $(\mathbb{GF}(p^m)^*, \times)$.

2.3.5.5. *Examples:* $\mathbb{F}_7$ *and* $\mathbb{F}_2[\xi]/\langle 1 + \xi + \xi^3 \rangle$

– Let $x = 3$ be an element of the field $\mathbb{GF}(7^1) = \mathbb{F}_7$. This element is a primitive element since it generates all the non-zero elements of $\mathbb{F}_7$. The element $x = 5$ is another primitive element of $\mathbb{F}_7$.

– Let α be a root of the polynomial $1 + \xi + \xi^3$. The elements of the field $\mathbb{GF}(2^3) = \mathbb{F}_2[\xi]/\langle 1 + \xi + \xi^3 \rangle$ are

$$0,\ \alpha,\ \alpha^2,\ 1+\alpha = \alpha^3,\ \alpha+\alpha^2 = \alpha^4$$

$$1+\alpha+\alpha^2 = \alpha^5,\ 1+\alpha^2 = \alpha^6,\ 1 = \alpha^7$$

so that α is a primitive element of $\mathbb{F}_2[\xi]/\langle 1 + \xi + \xi^3 \rangle$. Each of the elements $\alpha^2, \alpha^3, \cdots, \alpha^6$ is another primitive element of $\mathbb{F}_2[\xi]/\langle 1 + \xi + \xi^3 \rangle$.

2.3.5.6. *Number of primitive elements*

PROPOSITION 2.13.– The number of primitive elements of the Galois field $\mathbb{GF}(p^m)$ is $\varphi(p^m - 1)$, where φ is the Euler function (see Appendix for the definition of φ).

PROOF.– The proof follows from the isomorphism

$$(\mathbb{GF}(p^m)^*, \times) \simeq C_{p^m-1}$$

and the fact that the cyclic group C_n has $\varphi(n)$ generators. □

The order n of an element of the group $(\mathbb{GF}(p^m)^*, \times)$ is such that n divides $p^m - 1$. Therefore, if $p^m - 1$ is prime, then the field $\mathbb{GF}(p^m)$ admits $p^m - 2$ primitive elements. Furthermore, if $p^m - 1$ is prime, then each root of a prime polynomial $P_m(\xi)$ is a primitive element of the extension $\mathbb{GF}(p^m) = \mathbb{F}_p[\xi]/\langle P_m(\xi)\rangle$ and $P_m(\xi)$ is a primitive polynomial (see also 2.3.5.8).

2.3.5.7. *Examples:* $\mathbb{F}_7$ *and* $\mathbb{GF}(3^2)$

As the first example, the field $\mathbb{F}_7$ has $\varphi(7-1) = 2$ primitive elements, 3 and its inverse $3^{-1} = 5$; in detail

$$3^1 = 3, \quad 3^2 \equiv 2, \quad 3^3 \equiv 6, \quad 3^4 \equiv 4, \quad 3^5 \equiv 5, \quad 3^6 \equiv 1$$

$$(3^{-1})^1 = 5, \ (3^{-1})^2 \equiv 4, \ (3^{-1})^3 \equiv 6$$

$$(3^{-1})^4 \equiv 2, \ (3^{-1})^5 \equiv 3, \ (3^{-1})^6 \equiv 1$$

modulo 7.

As the second example, the field $\mathbb{GF}(3^2)$ has $\varphi(9-1) = 4$ primitive elements. Let us consider the realization $\mathbb{F}_3[\xi]/\langle 1+\xi^2\rangle$ of $\mathbb{GF}(3^2)$, and let α be a root of the prime polynomial $1+\xi^2$. Since

$$1+\alpha^2 = 0 \Rightarrow \alpha^2 = -1 \equiv 2 \Rightarrow \alpha^3 \equiv 2\alpha \Rightarrow \alpha^4 \equiv 2\alpha^2 \equiv 4 \equiv 1 \bmod 3$$

then α is not a primitive element of $\mathbb{F}_3[\xi]/\langle 1+\xi^2\rangle$. However, we can check that

$$a = 1+\alpha \Rightarrow a^2 \equiv 2\alpha, \quad a^3 \equiv 1+2\alpha, \quad a^4 \equiv 2$$

$$a^5 \equiv 2+2\alpha, \quad a^6 \equiv \alpha, \quad a^7 \equiv 2+\alpha, \quad a^8 \equiv 1 \bmod 3$$

so that $a = 1+\alpha$ is a primitive element of $\mathbb{F}_3[\xi]/\langle 1+\xi^2\rangle$; the inverse $a^{-1} = 2+\alpha$ of a is also a primitive element of $\mathbb{F}_3[\xi]/\langle 1+\xi^2\rangle$. The two other primitive elements are $b = 1+2\alpha$ and $b^{-1} = 2+2\alpha$.

2.3.5.8. *Primitive polynomial*

DEFINITION 2.5.– A prime polynomial $P_m(\xi)$ of degree m ($m \geq 1$) in $\mathbb{F}_p[\xi]$ for which a root α is a primitive element of the field $\mathbb{GF}(p^m) = \mathbb{F}_p[\xi]/\langle P_m(\xi)\rangle$ is called a *primitive polynomial* over $\mathbb{F}_p$ of degree m.

A primitive polynomial over $\mathbb{F}_p$ is irreducible over $\mathbb{F}_p$, but an irreducible polynomial is not necessarily primitive (there are irreducible polynomials that are not primitive).

2.3.5.9. *Counter-examples*

The polynomials $P_2(\xi) = 1 + \xi^2$ in $\mathbb{F}_3[\xi]$, $P_3(\xi) = 2 + 2\xi + \xi^3$ in $\mathbb{F}_3[\xi]$ and $P_4(\xi) = 1 + \xi + \xi^2 + \xi^3 + \xi^4$ in $\mathbb{F}_2[\xi]$ are irreducible but are not primitive. The non-primitive character of the three polynomials $P_2(\xi)$, $P_3(\xi)$ and $P_4(\xi)$ follows from the fact that their roots are not primitive elements of the fields $\mathbb{F}_3[\xi]/\langle P_2(\xi)\rangle$, $\mathbb{F}_3[\xi]/\langle P_3(\xi)\rangle$ and $\mathbb{F}_2[\xi]/\langle P_4(\xi)\rangle$, respectively.

2.3.5.10. *Examples*

We give in Tables 2.14 and 2.15 a list of primitive polynomials in $\mathbb{F}_2[\xi]$ and $\mathbb{F}_3[\xi]$, respectively.

For $m \geq 2$, all prime (i.e. monic + irreducible) polynomials $P_m(\xi)$ of degree m in $\mathbb{F}_p[\xi]$ are primitive polynomials if and only if $p^m - 1$ is prime. This explains for example that only two of the three prime polynomials $P_4(\xi)$ in $\mathbb{F}_2[\xi]$ are primitive polynomials: the prime polynomial $1 + \xi + \xi^2 + \xi^3 + \xi^4$ is not present in Table 2.14 (compare Tables 2.11 and 2.14).

Primitive Polynomial $P_m(\xi)$ in $\mathbb{F}_2[\xi]$	Degree m	$\mathbb{GF}(2^m)$
$1 + \xi$	1	$\mathbb{GF}(2)$
$1 + \xi + \xi^2$	2	$\mathbb{GF}(4)$
$1 + \xi + \xi^3$, $1 + \xi^2 + \xi^3$	3	$\mathbb{GF}(8)$
$1 + \xi + \xi^4$, $1 + \xi^3 + \xi^4$	4	$\mathbb{GF}(16)$
$1 + \xi^2 + \xi^5$, $1 + \xi^3 + \xi^5$ $1 + \xi + \xi^2 + \xi^3 + \xi^5$, $1 + \xi + \xi^2 + \xi^4 + \xi^5$ $1 + \xi + \xi^3 + \xi^4 + \xi^5$, $1 + \xi^2 + \xi^3 + \xi^4 + \xi^5$	5	$\mathbb{GF}(32)$
$1 + \xi + \xi^6$, $1 + \xi^5 + \xi^6$ $1 + \xi + \xi^2 + \xi^5 + \xi^6$, $1 + \xi + \xi^3 + \xi^4 + \xi^6$ $1 + \xi + \xi^4 + \xi^5 + \xi^6$, $1 + \xi^2 + \xi^3 + \xi^5 + \xi^6$	6	$\mathbb{GF}(64)$

Table 2.14. *Primitive polynomials $P_m(\xi)$ of degree m in $\mathbb{F}_2[\xi]$ for the field $\mathbb{GF}(2^m)$ with $1 \leq m \leq 6$*

Primitive Polynomial $P_m(\xi)$ in $\mathbb{F}_3[\xi]$	Degree m	$\mathbb{GF}(3^m)$
$1+\xi$	1	$\mathbb{GF}(3)$
$2+\xi+\xi^2,\quad 2+2\xi+\xi^2$	2	$\mathbb{GF}(9)$
$1+2\xi+\xi^3,\quad 1+2\xi^2+\xi^3$ $1+\xi+2\xi^2+\xi^3,\quad 1+2\xi+\xi^2+\xi^3$	3	$\mathbb{GF}(27)$
$2+\xi+\xi^4,\quad 2+2\xi+\xi^4$ $2+\xi^3+\xi^4,\quad 2+2\xi^3+\xi^4$ $2+\xi+\xi^2+2\xi^3+\xi^4,\quad 2+\xi+2\xi^2+2\xi^3+\xi^4$ $2+2\xi+\xi^2+\xi^3+\xi^4,\quad 2+2\xi+2\xi^2+\xi^3+\xi^4$	4	$\mathbb{GF}(81)$

Table 2.15. *Primitive polynomials $P_m(\xi)$ of degree m in $\mathbb{F}_3[\xi]$ for the field $\mathbb{GF}(3^m)$ with $1 \leq m \leq 4$*

2.3.5.11. *Existence*

PROPOSITION 2.14.– For any Galois field $\mathbb{GF}(p^m)$, there exists at least one primitive polynomial $P_m(\xi)$ of degree m over $\mathbb{F}_p$. The m roots of a primitive polynomial $P_m(\xi)$ over $\mathbb{F}_p$ are primitive elements of $\mathbb{GF}(p^m)$.

2.3.5.12. *Number of primitive polynomials*

PROPOSITION 2.15.– The number of primitive polynomials of the Galois field $\mathbb{GF}(p^m)$, i.e. the number of primitive polynomials of degree m over $\mathbb{F}_p$, is $\frac{1}{m}\varphi(p^m - 1)$, where φ is the Euler function (see Appendix for the definition of φ).

Table 2.16 gives the number of primitive elements and primitive polynomials for some fields $\mathbb{GF}(p^m)$.

p^m	2^1	3^1	2^2	5^1	7^1	2^3	3^2	2^4	3^3	2^5	2^6	3^4
$\varphi(p^m-1)$	1	1	2	2	2	6	4	8	12	30	36	32
$\frac{1}{m}\varphi(p^m-1)$	1	1	1	2	2	2	2	2	4	6	6	8

Table 2.16. *Number of primitive elements $\varphi(p^m-1)$ and primitive polynomials $\frac{1}{m}\varphi(p^m-1)$ for some Galois fields $\mathbb{GF}(p^m)$*

2.3.6. *Logarithm of a field element*

We have seen that any non-zero element x of $\mathbb{GF}(p^m)$ can be written as $x = \alpha^s$ with $s = 0, 1, \cdots, p^m - 2$ where α is a generator of the group $(\mathbb{GF}(p^m)^*, \times)$. This yields the following definition.

DEFINITION 2.6.– The power s in the non-zero element $x = \alpha^s$ $(s = 0, 1, \cdots, p^m - 2)$ of $\mathbb{GF}(p^m)$, where α is a primitive element of $\mathbb{GF}(p^m)$, is called the *discrete logarithm* of x in the α logarithmic basis. In other words,

$$\forall x \in \mathbb{GF}(p^m)^* : x = \alpha^s, \; s \in \{0, 1, \cdots, p^m - 2\} \Rightarrow \log_\alpha(x) = s$$

where $\log_\alpha(x)$ is the discrete logarithm of x in the α logarithmic basis.

Note that:

– In view of $\alpha^{p^m-1} = 1$, the discrete logarithm is defined modulo $p^m - 1$.

– The knowledge of the non-zero elements of a field $\mathbb{GF}(p^m)$ as powers of a primitive element α renders possible, via the use of the discrete logarithm, to calculate the product $z = xy$ of two arbitrary non-zero elements x and y of $\mathbb{GF}(p^m)$. By example, for $\mathbb{GF}(2^3) = \mathbb{F}_2[\xi]/\langle 1 + \xi + \xi^3 \rangle$, we have seen that the non-zero elements are (see 2.3.5.5)

$$\alpha, \; \alpha^2, \; 1 + \alpha = \alpha^3, \; \alpha + \alpha^2 = \alpha^4, \; 1 + \alpha + \alpha^2 = \alpha^5$$

$$1 + \alpha^2 = \alpha^6, \; 1 = \alpha^7$$

Therefore,

$$x = 1 + \alpha + \alpha^2 = \alpha^5 \Rightarrow \log_\alpha(x) = 5$$

$$y = 1 + \alpha^2 = \alpha^6 \Rightarrow \log_\alpha(y) = 6$$

$$\Rightarrow \log_\alpha(xy) = 5 + 6 = 11 \equiv 4 \bmod 7 \; \Rightarrow \; xy = \alpha^4 = \alpha + \alpha^2$$

where we have used the fact that here the discrete logarithm is defined modulo 7.

2.3.7. *Practical rules for constructing a Galois field*

The general procedure for constructing the field $\mathbb{GF}(p^m)$, with p^m elements (p prime and $m \geq 2$), from the base field $\mathbb{F}_p$, with p elements, is based on the following rules.

– To look for a monic polynomial $P_m(\xi)$ of degree m, with coefficients in $\mathbb{F}_p$ and irreducible in $\mathbb{F}_p[\xi]$. (For every prime p, there exists at least one monic irreducible polynomial $P_m(\xi)$ with $m \geq 1$.) More precisely, we need to find a non-zero polynomial

$$P_m(\xi) = c_0 + c_1\xi + \cdots + c_{m-1}\xi^{m-1} + \xi^m, \quad c_0, c_1, \cdots, c_{m-1} \in \mathbb{F}_p$$

that cannot be factored into the product of two polynomials, of degree lower than m, in $\mathbb{F}_p[\xi]$.

In general, the polynomial $P_m(\xi)$ is not unique. The multiplication table of the elements of $\mathbb{GF}(p^m)$ depends on the choice of $P_m(\xi)$. However, all the choices made for $P_m(\xi)$ yield isomorphic realizations, that is to say to the same field $\mathbb{GF}(p^m)$.

Note that the polynomial $P_m(\xi)$ in $\mathbb{F}_p[\xi]$ can also be considered as a polynomial in $\mathbb{GF}(p^m)[\xi]$.

– To consider a root α of $P_m(\xi)$ (it is unnecessary to exhibit the detailed expression of α). This root is an element of $\mathbb{GF}(p^m)$ which does not belong to $\mathbb{F}_p$. Up to this point, one has $p + 1$ elements of $\mathbb{GF}(p^m)$: p elements from $\mathbb{F}_p$ and the element α.

– The other elements of the field $\mathbb{GF}(p^m)$ can be represented by polynomials in α of degree lower than or equal to $m - 1$ with coefficients in $\mathbb{F}_p$. Indeed, every element x of $\mathbb{GF}(p^m)$ can be expressed in the form

$$x \text{ or } x(\alpha) = \sum_{k=0}^{m-1} x_k\alpha^k = x_0 + x_1\alpha + \cdots + x_{m-1}\alpha^{m-1}$$

with the coefficients $x_0, x_1, \cdots, x_{m-1}$ in the field $\mathbb{F}_p$. Note that the element $x_0 + x_1\alpha + \cdots + x_m\alpha^{m-1}$ can be considered as a vector of components $[x_0\, x_1\, \cdots\, x_{m-1}]$ or $[x_0\, x_1\, \cdots\, x_{m-1}]_\alpha$ in an m-dimensional vector space over the field $\mathbb{F}_p$. It is easily seen that there are p^m expressions of type $\sum_{k=0}^{m-1} x_k\alpha^k$ giving the p^m elements of $\mathbb{GF}(p^m)$; the elements of $\mathbb{F}_p$ (viz. $[x_0\, 0\, \cdots\, 0]$) and α (viz. $[0\, 1\, \cdots\, 0]$) are particular cases of $\sum_{k=0}^{m-1} x_k\alpha^k$.

As a résumé, the so-obtained Galois field $\mathbb{F}_p[\xi]/\langle P_m(\xi)\rangle$ is the unique (up to isomorphism) extension of degree m of the base field $\mathbb{F}_p$ by the element α, a root of the prime polynomial $P_m(\xi)$ (α is the residue class of ξ modulo $P_m(\xi)$). It is convenient to use the notation $\mathbb{F}_p[\xi]/\langle P_m(\xi)\rangle$ for describing the field $\mathbb{GF}(p^m)$. The p^m elements of $\mathbb{GF}(p^m)$ are represented by residue classes of polynomials in $\mathbb{F}_p[\xi]$. The residue classes are obtained by effecting the relevant additions and multiplications modulo p and modulo $P_m(\alpha) = 0$. The addition of elements of $\mathbb{GF}(p^m)$ is that of vectors in a vector space over $\mathbb{F}_p$. The product of elements is the remainder of the division by $P_m(\xi)$ of the product in $\mathbb{F}_p[\xi]$. Indeed, all the calculations in $\mathbb{GF}(p^m)$ are made by using $P_m(\alpha) = 0$ and $px = 0$ for any x in $\mathbb{GF}(p^m)$. In this regard, let us suppose that, in a calculation, an element α^k appears with $k \geq m$. Then, the k power of α can be decreased by repeated use of

$$\alpha^m = -(c_0 + c_1\alpha + \cdots + c_{m-1}\alpha^{m-1})$$

(that corresponds to $P_m(\alpha) = 0$) in

$$\alpha^k = \alpha^{k-m}\alpha^m$$

and of

$$p \times y = 0$$

where y is any positive power of α. Finally, intermediate calculations and the realization of the elements of $\mathbb{F}_p[\xi]/\langle P_m(\xi)\rangle$ depend on $P_m(\xi)$. However, for fixed p, all the possible choices of $P_m(\xi)$ of the same degree m give isomorphic realizations of $\mathbb{GF}(p^m)$.

The limiting case $m = 1$ deserves special attention. For example, let us consider $\mathbb{GF}(2) = \mathbb{F}_2[\xi]/\langle P_1(\xi)\rangle$. There are two prime polynomials $P_1(\xi)$ of degree 1 in $\mathbb{F}_2[\xi]$, namely, ξ and $1 + \xi$. The elements of $\mathbb{F}_2[\xi]/\langle\xi\rangle$ as well as of $\mathbb{F}_2[\xi]/\langle 1 + \xi\rangle$ are then x_0 where $x_0 \in \mathbb{F}_2$ so that $\mathbb{GF}(2)$ is the field $\mathbb{F}_2$. Similarly, it is easy to show that $\mathbb{GF}(p) = \mathbb{F}_p[\xi]/\langle P_1(\xi)\rangle$ can be identified as the field $\mathbb{F}_p$.

2.3.8. *Examples of extensions of fields*

Most of the examples below are devoted to Galois fields (one of the main subjects of this book). However, we start with two examples concerned with infinite fields.

2.3.8.1. *Example:* $\mathbb{Q}(\mathrm{i}) = \mathbb{Q}[\xi]/\langle 1 + \xi^2 \rangle$

It is easily checked that $\mathbb{Q}(\mathrm{i}) = \{a + ib \mid a, b \in \mathbb{Q}\}$ is an infinite field with respect to the addition and multiplication of complex numbers. The polynomial $1 + \xi^2$ is irreducible over $\mathbb{Q}$. Therefore, $\mathbb{Q}[\xi]/\langle 1 + \xi^2 \rangle$ is a field. This field is isomorphic to $\mathbb{Q}(\mathrm{i})$.

2.3.8.2. *Example:* $\mathbb{C} = \mathbb{R}[\xi]/\langle 1 + \xi^2 \rangle$

The polynomial $1 + \xi^2$ is irreducible over $\mathbb{R}$. Therefore, $\mathbb{R}[\xi]/\langle 1 + \xi^2 \rangle$ is a field whose elements are $x_0 + x_1\alpha$, where x_0 and x_1 are in $\mathbb{R}$, and α is a root of $1 + \xi^2 = 0$. It is clear that this infinite field is isomorphic to the field $\mathbb{C}$ of complex numbers.

More generally, let us consider the polynomial $c_0 + c_1\xi + \xi^2$ with $c_1^2 - 4c_0 < 0$ (the above example corresponds to $c_0 - 1 = c_1 = 0$). This polynomial is irreducible over $\mathbb{R}$. Thus, $\mathbb{R}[\xi]/\langle c_0 + c_1\xi + \xi^2 \rangle$ is a field. This field is isomorphic to $\mathbb{C}$.

2.3.8.3. *Example:* $\mathbb{GF}(2^2) = \mathbb{F}_2[\xi]/\langle 1 + \xi + \xi^2 \rangle$

This example is treated in detail in section 2.2. Here, $p = 2$ and $m = 2$. In this case, the elements of $\mathbb{GF}(2^2)$ are

$$0, \ 1, \ \alpha, \ 1 + \alpha \equiv \alpha^2 \bmod 2$$

where α is a root of the irreducible polynomial

$$P_2(\xi) = 1 + \xi + \xi^2$$

belonging to $\mathbb{F}_2[\xi]$. Tables 2.6 (or 2.8) and 2.7 (or 2.9) give the addition and multiplication tables of $\mathbb{GF}(2^2)$, respectively. Note that $\alpha^3 \equiv 1 \bmod 2$ ($\Rightarrow \alpha^4 \equiv \alpha \bmod 2$). Therefore, the elements of $\mathbb{GF}(2^2)$ can be taken in the form 0, α^1, α^2 and α^3 (according to section 2.3.5, α is said to be a primitive element). As already mentioned in section 2.3.4.2, the additive group $(\mathbb{GF}(2^2), +)$ is isomorphic to the direct product $C_2 \times C_2$ (isomorphic to the Klein four-group V) and the multiplicative group $(\mathbb{GF}(2^2)^*, \times)$ is isomorphic to the cyclic group C_3 (isomorphic to the group $(\mathbb{F}_3, +)$).

2.3.8.4. *Example:* $\mathbb{GF}(2^3) = \mathbb{F}_2[\xi]/\langle 1 + \xi + \xi^3 \rangle$

In this pedagogical example, $p = 2$ and $m = 3$ ($\Rightarrow p^m = 8$). Therefore, each monic irreducible (i.e. prime) polynomial $P_3(\xi)$ in $\mathbb{F}_2[\xi]$ is of the form

$$P_3(\xi) = c_0 + c_1\xi + c_2\xi^2 + \xi^3$$

where c_0, c_1 and c_2 are in $\mathbb{F}_2$. We can take

$$P_3(\xi) = 1 + \xi + \xi^3$$

Let α be a root of $P_3(\xi)$. The elements x of $\mathbb{GF}(2^3)$ are of the type

$$x = x_0 + x_1\alpha + x_2\alpha^2$$

(here $m - 1 = 2$) where x_0, x_1 and x_2 belong to $\mathbb{F}_2$. Alternatively, we can write x as

$$x = [x_0x_1x_2]_\alpha \text{ or simply } [x_0x_1x_2]$$

Therefore, the elements of $\mathbb{GF}(2^3)$ are

$$0 = [000], \quad \alpha = [010], \quad \alpha^2 = [001], \quad 1 + \alpha = [110]$$

$$\alpha + \alpha^2 = [011], \quad 1 + \alpha + \alpha^2 = [111], \quad 1 + \alpha^2 = [101], \quad 1 = [100]$$

By taking into account

$$P_3(\alpha) = 0 \Rightarrow 1 + \alpha + \alpha^3 = 0$$

we obtain

$$\alpha^3 \equiv 1 + \alpha, \quad \alpha^4 \equiv \alpha + \alpha^2, \quad \alpha^5 \equiv 1 + \alpha + \alpha^2$$

$$\alpha^6 \equiv 1 + \alpha^2, \quad \alpha^7 \equiv 1$$

modulo 2. As a résumé, the $2^3 = 8$ elements 0, a, b, c, d, e, f and 1 of $\mathbb{GF}(2^3) = \mathbb{F}_2[\xi]/\langle 1 + \xi + \xi^3 \rangle$ read

$$0, \quad a = \alpha, \quad b = \alpha^2, \quad c = 1 + \alpha \equiv \alpha^3, \quad d = \alpha + \alpha^2 \equiv \alpha^4$$

$$e = 1 + \alpha + \alpha^2 \equiv \alpha^5, \quad f = 1 + \alpha^2 \equiv \alpha^6, \quad 1 \equiv \alpha^7$$

modulo 2 and by taking into account $P_3(\alpha) = 0$. Clearly, the element α is of order 7 in the group $(\mathbb{GF}(2^3)^*, \times)$.

From a formal point of view, the addition and multiplication of two elements $x = x_0 + x_1\alpha + x_2\alpha^2$ and $y = y_0 + y_1\alpha + y_2\alpha^2$ are given by

$$\begin{aligned} x + y &= x_0 + x_1\alpha + x_2\alpha^2 + y_0 + y_1\alpha + y_2\alpha^2 \\ &= x_0 + y_0 + (x_1 + y_1)\alpha + (x_2 + y_2)\alpha^2 \end{aligned}$$

and

$$\begin{aligned} x \times y &= (x_0 + x_1\alpha + x_2\alpha^2) \times (y_0 + y_1\alpha + y_2\alpha^2) \\ &= x_0y_0 + (x_1y_0 + x_0y_1)\alpha + (x_0y_2 + x_2y_0 + x_1y_1)\alpha^2 \\ &\quad + (x_1y_2 + x_2y_1)\alpha^3 + x_2y_2\alpha^4 \end{aligned}$$

By making use of $\alpha^3 \equiv 1 + \alpha$ and $\alpha^4 \equiv \alpha + \alpha^2$ modulo 2, we have

$$\begin{aligned} x \times y \equiv\ & x_0y_0 + x_1y_2 + x_2y_1 + (x_1y_0 + x_0y_1 + x_1y_2 + x_2y_1 + x_2y_2)\alpha \\ & +(x_0y_2 + x_2y_0 + x_1y_1 + x_2y_2)\alpha^2 \end{aligned}$$

modulo 2.

From a practical point of view, it is generally simpler to directly calculate $x + y$ and $x \times y$ rather than using the preceding formulas. As an example of calculation, we have

$$c + f = 1 + \alpha + 1 + \alpha^2 \equiv \alpha + \alpha^2 = d \bmod 2$$

and

$$\begin{aligned} & c \times f = (1 + \alpha) \times (1 + \alpha^2) = 1 + \alpha^2 + \alpha + \alpha^3 \\ & P_3(\alpha) = 0 \Rightarrow \alpha^3 = -1 - \alpha \equiv 1 + \alpha \bmod 2 \\ & \Rightarrow c \times f \equiv 1 + \alpha^2 + \alpha + 1 + \alpha \equiv \alpha^2 = b \bmod 2 \end{aligned}$$

Repeated calculations of this type yield the addition and multiplication tables of $\mathbb{GF}(2^3) = \mathbb{F}_2[\xi]/\langle 1+\xi+\xi^3\rangle$, see Tables 2.17 and 2.18. Of course, the obtained tables are symmetrical with respect to the diagonal since the laws $+$ and $\times$ are commutative.

$+$	0	$1{=}\alpha^7$	$a{=}\alpha^1$	$b{=}\alpha^2$	$c{=}\alpha^3$	$d{=}\alpha^4$	$e{=}\alpha^5$	$f{=}\alpha^6$
0	0	1	a	b	c	d	e	f
$1{=}\alpha^7$		0	c	f	a	e	d	b
$a{=}\alpha^1$			0	d	1	b	f	e
$b{=}\alpha^2$				0	e	a	c	1
$c{=}\alpha^3$					0	f	b	d
$d{=}\alpha^4$						0	1	c
$e{=}\alpha^5$							0	a
$f{=}\alpha^6$								0

Table 2.17. *Addition table for $\mathbb{GF}(2^3) = \mathbb{F}_2[\xi]/\langle 1+\xi+\xi^3\rangle$ where $a=\alpha$, $b=\alpha^2$, $c=1+\alpha\equiv\alpha^3$, $d=\alpha+\alpha^2\equiv\alpha^4$, $e=1+\alpha+\alpha^2\equiv\alpha^5$, $f=1+\alpha^2\equiv\alpha^6$ and $1\equiv\alpha^7$ where α is a primitive element (solution of $1+\xi+\xi^3=0$); the element at the intersection of the line x and the column y is $x+y$ (the elements below the diagonal of the table are obtained by using $x+y=y+x$)*

$\times$	0	$1{=}\alpha^7$	$a{=}\alpha^1$	$b{=}\alpha^2$	$c{=}\alpha^3$	$d{=}\alpha^4$	$e{=}\alpha^5$	$f{=}\alpha^6$
0	0	0	0	0	0	0	0	0
$1{=}\alpha^7$		1	a	b	c	d	e	f
$a{=}\alpha^1$			b	c	d	e	f	1
$b{=}\alpha^2$				d	e	f	1	a
$c{=}\alpha^3$					f	1	a	b
$d{=}\alpha^4$						a	b	c
$e{=}\alpha^5$							c	d
$f{=}\alpha^6$								e

Table 2.18. *Multiplication table for $\mathbb{GF}(2^3) = \mathbb{F}_2[\xi]/\langle 1+\xi+\xi^3\rangle$ where $a=\alpha$, $b=\alpha^2$, $c=1+\alpha\equiv\alpha^3$, $d=\alpha+\alpha^2\equiv\alpha^4$, $e=1+\alpha+\alpha^2\equiv\alpha^5$, $f=1+\alpha^2\equiv\alpha^6$ and $1\equiv\alpha^7$ where α is a primitive element (solution of $1+\xi+\xi^3=0$); the element at the intersection of the line x and the column y is $x\times y$ (the elements below the diagonal of the table are obtained by using $x\times y=y\times x$)*

Note that another way to obtain the expression of α^s as a polynomial in α of degree lower than or equal to 2 (for $s = 1$ to 7) is to directly calculate the corresponding residue classes modulo $P_3(\alpha)=0$. For instance, the Euclidean division of ξ^6 by $P_3(\xi)$ yields

$$\xi^6 = P_3(\xi)\times(\xi^3-\xi-1)+(1+2\xi+\xi^2)$$

Therefore, for $\xi = \alpha$, we obtain

$$\alpha^6 = 1 + 2\alpha + \alpha^2 \equiv 1 + \alpha^2 \bmod 2$$

The remaining α^s (for $s = 3, 4, 5, 7$) can be obtained in the same way.

As far as group theory is concerned, the two following results should be emphasized. First, the multiplicative group $(\mathbb{GF}(2^3)^*, \times)$ is isomorphic to the cyclic group C_7 of order 7 and α is a generator of the group $(\mathbb{GF}(2^3)^*, \times)$ consisting of the elements $\alpha^1, \alpha^2, \cdots, \alpha^7$. Hence, α is a primitive element and $1 + \xi + \xi^3$ a primitive polynomial in $\mathbb{F}_2[\xi]$ (see section 2.3.5). Second, the additive group $(\mathbb{GF}(2^3), +)$ is isomorphic to the direct product $C_2 \times C_2 \times C_2$ of order 8.

To end up with this example, let us note that

$$\alpha^7 \equiv 1 \Rightarrow \alpha^8 \equiv \alpha$$

and that $\mathbb{F}_2$ is a sub-field of $\mathbb{GF}(2^3)$.

2.3.8.5. *Example:* $\mathbb{GF}(2^3) = \mathbb{F}_2[\xi]/\langle 1 + \xi^2 + \xi^3 \rangle$

To construct $\mathbb{GF}(2^3)$ as an extension of $\mathbb{F}_2$, we can take

$$P_3(\xi) = 1 + \xi^2 + \xi^3$$

as a prime polynomial in $\mathbb{F}_2[\xi]$ instead of $P_3(\xi) = 1+\xi+\xi^3$ as in the preceding example. Let β be a root of $P_3(\xi)$. The elements of $\mathbb{GF}(2^3) = \mathbb{F}_2[\xi]/\langle 1+\xi^2+\xi^3 \rangle$ are then

$$0, \quad a' = 1 + \beta^2 \equiv \beta^3, \quad b' = \beta + \beta^2 \equiv \beta^6, \quad c' = \beta^2$$
$$d' = 1 + \beta \equiv \beta^5, \quad e' = \beta, \quad f' = 1 + \beta + \beta^2 \equiv \beta^4, \quad 1 \equiv \beta^7$$

modulo 2 and modulo $P_3(\beta) = 0$. Each non-zero element of $\mathbb{F}_2[\xi]/\langle 1 + \xi^2 + \xi^3 \rangle$ is a power β^s ($s = 1$ to 7) of β so that β is a primitive element and $1 + \xi^2 + \xi^3$ a primitive polynomial in $\mathbb{F}_2[\xi]$. The addition and multiplication tables of $\mathbb{GF}(2^3) = \mathbb{F}_2[\xi]/\langle 1 + \xi^2 + \xi^3 \rangle$ are given in Tables 2.19 and 2.20, respectively.

$+$	0	$1{=}\beta^7$	$a'{=}\beta^3$	$b'{=}\beta^6$	$c'{=}\beta^2$	$d'{=}\beta^5$	$e'{=}\beta^1$	$f'{=}\beta^4$
0	0	1	a'	b'	c'	d'	e'	f'
$1{=}\beta^7$		0	c'	f'	a'	e'	d'	b'
$a'{=}\beta^3$			0	d'	1	b'	f'	e'
$b'{=}\beta^6$				0	e'	a'	c'	1
$c'{=}\beta^2$					0	f'	b'	d'
$d'{=}\beta^5$						0	1	c'
$e'{=}\beta^1$							0	a'
$f'{=}\beta^4$								0

Table 2.19. *Addition table for* $\mathbb{GF}(2^3) = \mathbb{F}_2[\xi]/\langle 1 + \xi^2 + \xi^3\rangle$ *where* $a' = 1 + \beta^2 \equiv \beta^3$, $b' = \beta + \beta^2 \equiv \beta^6$, $c' = \beta^2$, $d' = 1 + \beta \equiv \beta^5$, $e' = \beta$, $f' = 1 + \beta + \beta^2 \equiv \beta^4$ *and* $1 \equiv \beta^7$ *where* β *is a primitive element, solution of* $1 + \xi^2 + \xi^3 = 0$*; this table coincides with Table 2.17 through the correspondence* $a \leftrightarrow a'$, $b \leftrightarrow b'$, $\cdots$, $f \leftrightarrow f'$

$\times$	0	$1{=}\beta^7$	$a'{=}\beta^3$	$b'{=}\beta^6$	$c'{=}\beta^2$	$d'{=}\beta^5$	$e'{=}\beta^1$	$f'{=}\beta^4$
0	0	0	0	0	0	0	0	0
$1{=}\beta^7$		1	a'	b'	c'	d'	e'	f'
$a'{=}\beta^3$			b'	c'	d'	e'	f'	1
$b'{=}\beta^6$				d'	e'	f'	1	a'
$c'{=}\beta^2$					f'	1	a'	b'
$d'{=}\beta^5$						a'	b'	c'
$e'{=}\beta^1$							c'	d'
$f'{=}\beta^4$								e'

Table 2.20. *Multiplication table for* $\mathbb{GF}(2^3) = \mathbb{F}_2[\xi]/\langle 1+\xi^2+\xi^3\rangle$ *where* $a' = 1+\beta^2 \equiv \beta^3$, $b' = \beta + \beta^2 \equiv \beta^6$, $c' = \beta^2$, $d' = 1 + \beta \equiv \beta^5$, $e' = \beta$, $f' = 1 + \beta + \beta^2 \equiv \beta^4$ *and* $1 \equiv \beta^7$ *where* β *is a primitive element, solution of* $1 + \xi^2 + \xi^3 = 0$*; this table coincides with Table 2.18 through the correspondence* $a \leftrightarrow a'$, $b \leftrightarrow b'$, $\cdots$, $f \leftrightarrow f'$

A comparison between $\mathbb{F}_2[\xi]/\langle 1 + \xi^2 + \xi^3\rangle$ and $\mathbb{F}_2[\xi]/\langle 1 + \xi + \xi^3\rangle$ (see 2.3.8.4) is in order. It is a simple matter of calculation to verify that the bijection defined by

$$0 \leftrightarrow 0, \quad 1 \leftrightarrow 1, \quad a \leftrightarrow a', \quad b \leftrightarrow b', \quad \cdots, \quad f \leftrightarrow f'$$

shows that the two fields $\mathbb{F}_2[\xi]/\langle 1 + \xi + \xi^3\rangle$ and $\mathbb{F}_2[\xi]/\langle 1 + \xi^2 + \xi^3\rangle$ are isomorphic (the elements $a, b, \cdots, f$ are defined in 2.3.8.4; the addition and multiplication tables in terms of β coincide with those in terms of α through the latter one-to-one correspondence). It can be checked that the introduction of

$$\beta = 1 + \alpha + \alpha^2$$

into $1 + \beta^2 + \beta^3 = 0$ yields $1 + \alpha + \alpha^3 = 0$ modulo 2 (α is a root of $1 + \xi + \xi^3 = 0$; see 2.3.8.4).

Examples 2.3.8.4 and 2.3.8.5 emphasize the fact that, for p and m fixed, the field $\mathbb{GF}(p^m)$ is unique (up to an isomorphism). They also show that the sole notation $\mathbb{GF}(p^m)$ is not sufficient to give an account of the expression (in terms of a primitive element) of the elements of a field of cardinal p^m. Indeed, the expression of the element $x_0 + x_1\alpha + \cdots + x_{m-1}\alpha^{m-1}$ in terms of a power α^s of α (with $s = m, m+1, \cdots, p^m - 2$) depends on the chosen monic irreducible polynomial. However, two realizations of the same field $\mathbb{GF}(p^m)$ corresponding to two distinct prime polynomials are isomorphic.

2.3.8.6. *Example:* $\mathbb{GF}(2^4) = \mathbb{F}_2[\xi]/\langle 1 + \xi + \xi^4 \rangle$

For the field $\mathbb{GF}(2^4)$, we have $p = 2$ and $m = 4$. We can take

$$P_4(\xi) = 1 + \xi + \xi^4$$

as monic irreducible polynomial in $\mathbb{F}_2[\xi]$. The 16 elements x of the field $\mathbb{GF}(2^4) = \mathbb{F}_2[\xi]/\langle 1 + \xi + \xi^4 \rangle$ are then of the form

$$x = \sum_{k=0}^{m-1=3} x_k \alpha^k = x_0 + x_1\alpha + x_2\alpha^2 + x_3\alpha^3$$

where α is a solution of $P_4(\xi) = 0$ and x_0, x_1, x_2 and x_3 are elements of the field $\mathbb{F}_2$. (In passing, note that x looks like an element of a vector space of dimension $m = 4$ over the field $\mathbb{F}_p$ with $p = 2$.) The elements x of the field $\mathbb{GF}(2^4)$ are easily determined to be

$$0, \quad \alpha, \quad \alpha^2, \quad \alpha^3, \quad 1 + \alpha \equiv \alpha^4, \quad \alpha + \alpha^2 \equiv \alpha^5$$

$$\alpha^2 + \alpha^3 \equiv \alpha^6, \quad 1 + \alpha + \alpha^3 \equiv \alpha^7, \quad 1 + \alpha^2 \equiv \alpha^8$$

$$\alpha + \alpha^3 \equiv \alpha^9, \quad 1 + \alpha + \alpha^2 \equiv \alpha^{10}, \quad \alpha + \alpha^2 + \alpha^3 \equiv \alpha^{11}$$

$$1 + \alpha + \alpha^2 + \alpha^3 \equiv \alpha^{12}, \quad 1 + \alpha^2 + \alpha^3 \equiv \alpha^{13}$$

$$1 + \alpha^3 \equiv \alpha^{14}, \quad 1 \equiv \alpha^{15}$$

modulo $P_4(\alpha) = 0$ and modulo 2. From these expressions, we can easily set up the addition and multiplication tables of the field $\mathbb{GF}(2^4) = \mathbb{F}_2[\xi]/\langle 1+\xi+\xi^4 \rangle$. By introducing the notation $[x_0\ x_1\ x_2\ x_3]$ to denote the element $x_0 + x_1\alpha + x_2\alpha^2 + x_3\alpha^3$, we have

$$0 = [0000], \quad \alpha = [0100], \quad \alpha^2 = [0010], \ \alpha^3 = [0001], \ \alpha^4 = [1100]$$

$$\alpha^5 = [0110], \quad \alpha^6 = [0011], \quad \alpha^7 = [1101], \quad \alpha^8 = [1010]$$

$$\alpha^9 = [0101], \quad \alpha^{10} = [1110], \quad \alpha^{11} = [0111]$$

$$\alpha^{12} = [1111], \quad \alpha^{13} = [1011], \quad \alpha^{14} = [1001], \quad \alpha^{15} = [1000]$$

Note that α is a primitive element of $\mathbb{GF}(2^4) = \mathbb{F}_2[\xi]/\langle 1 + \xi + \xi^4 \rangle$ since it generates the multiplicative group $(\mathbb{GF}(2^4)^*, \times)$ isomorphic to the cyclic group C_{15}. Therefore, $1 + \xi + \xi^4$ is a primitive polynomial.

The other possible choices

$$1 + \xi^3 + \xi^4, \quad 1 + \xi + \xi^2 + \xi^3 + \xi^4$$

for $P_4(\xi)$ yield the same field (up to an isomorphism). Let us examine in turn the two choices.

– The polynomial $1 + \xi^3 + \xi^4$ is a primitive polynomial because if β is a root of $1 + \xi^3 + \xi^4 = 0$, then the elements of $\mathbb{F}_2[\xi]/\langle 1 + \xi^3 + \xi^4 \rangle$ are

$$0, \quad \beta, \quad \beta^2, \quad \beta^3, \quad 1 + \beta^3 \equiv \beta^4, \quad 1 + \beta + \beta^3 \equiv \beta^5$$

$$1 + \beta + \beta^2 + \beta^3 \equiv \beta^6, \quad 1 + \beta + \beta^2 \equiv \beta^7, \quad \beta + \beta^2 + \beta^3 \equiv \beta^8$$

$$1 + \beta^2 \equiv \beta^9, \quad \beta + \beta^3 \equiv \beta^{10}, \quad 1 + \beta^2 + \beta^3 \equiv \beta^{11}$$

$$1 + \beta \equiv \beta^{12}, \quad \beta + \beta^2 \equiv \beta^{13}, \quad \beta^2 + \beta^3 \equiv \beta^{14}, \quad 1 \equiv \beta^{15}$$

modulo $1 + \beta^3 + \beta^4 = 0$ and modulo 2 (the smallest positive integer n such that $\beta^n = 1$ is $n = 2^4 - 1$).

– The polynomial $1+\xi+\xi^2+\xi^3+\xi^4$ is not a primitive polynomial because if γ is a root of $1 + \xi + \xi^2 + \xi^3 + \xi^4 = 0$, then the smallest positive integer n such that $\gamma^n = 1$ is $n = 5 < 2^4 - 1$, modulo $1 + \gamma + \gamma^2 + \gamma^3 + \gamma^4 = 0$ and modulo 2.

2.3.8.7. *Example:* $\mathbb{GF}(3^2)$

In this case, we have $p = 3$ and $m = 2$. The nine distinct elements $x = x_0 + x_1\alpha$ of the field $\mathbb{GF}(3^2)$ represented by polynomials in α of degree lower or equal to 1 are

$$x = 0,\ 1,\ 2,\ \alpha,\ 1+\alpha,\ 2+\alpha,\ 2\alpha,\ 1+2\alpha,\ 2+2\alpha$$

where α is a root of a prime polynomial $P_2(\xi)$. The latter expressions for x and the addition table for the field $\mathbb{GF}(3^2)$ are independent of the chosen prime polynomial. We give in Table 2.21 the addition table for the group $(\mathbb{GF}(3^2), +)$ isomorphic to the direct product $C_3 \times C_3$: in view of the isomorphism of C_3 onto $(\mathbb{F}_3, +)$, the isomorphism $(\mathbb{GF}(3^2), +) \to C_3 \times C_3$ is described by the following correspondences

$$0 \leftrightarrow [00], \quad 1 \leftrightarrow [10], \quad 2 \leftrightarrow [20], \quad \alpha \leftrightarrow [01], \quad 1+\alpha \leftrightarrow [11]$$
$$2+\alpha \leftrightarrow [21], \quad 2\alpha \leftrightarrow [02], \quad 1+2\alpha \leftrightarrow [12], \quad 2+2\alpha \leftrightarrow [22]$$

where $[ab]$ is used to denote an element of $C_3 \times C_3$ (with a in the first C_3 and b in the second C_3).

+	[00]	[10]	[20]	[01]	[11]	[21]	[02]	[12]	[22]
[00]	[00]	[10]	[20]	[01]	[11]	[21]	[02]	[12]	[22]
[10]		[20]	[00]	[11]	[21]	[01]	[12]	[22]	[02]
[20]			[10]	[21]	[01]	[11]	[22]	[02]	[12]
[01]				[02]	[12]	[22]	[00]	[10]	[20]
[11]					[22]	[02]	[10]	[20]	[00]
[21]						[12]	[20]	[00]	[10]
[02]							[01]	[11]	[21]
[12]								[21]	[01]
[22]									[11]

Table 2.21. *Addition table for $\mathbb{GF}(3^2)$ where $[00] \leftrightarrow 0$, $[10] \leftrightarrow 1$, $[20] \leftrightarrow 2$, $[01] \leftrightarrow \alpha$, $[11] \leftrightarrow 1+\alpha$, $[21] \leftrightarrow 2+\alpha$, $[02] \leftrightarrow 2\alpha$, $[12] \leftrightarrow 1+2\alpha$ and $[22] \leftrightarrow 2+2\alpha$ (α is a solution of any prime polynomial $P_2(\xi)$ over $\mathbb{F}_3$, and $[ab]$ is used to denote an element of the group $(\mathbb{F}_3, +) \times (\mathbb{F}_3, +)$ isomorphic to $C_3 \times C_3$); the element at the intersection of the line x and the column y is $x + y$ (the elements below the diagonal of the table are obtained by using $x + y = y + x$)*

On the contrary, the expressions of the elements x of $\mathbb{GF}(3^2)$ in terms of powers of α do depend on the chosen prime polynomial. We can make three choices for the prime polynomial $P_2(\xi)$.

– The case $\mathbb{GF}(3^2) = \mathbb{F}_3[\xi]/\langle 1 + \xi^2 \rangle$. Here, we take

$$P_2(\xi) = 1 + \xi^2$$

as a prime polynomial in $\mathbb{F}_3[\xi]$. Then, the elements of $\mathbb{GF}(3^2)$ are

$$0, \quad 1 = \alpha^4 = \alpha^8, \quad 2 = \alpha^2 = \alpha^6, \quad \alpha = \alpha^5 = \alpha^9$$

$$1 + \alpha = \alpha + \alpha^4, \quad 2 + \alpha = \alpha + \alpha^2, \quad 2\alpha = \alpha^3 = \alpha^7$$

$$1 + 2\alpha = \alpha^3 + \alpha^4, \quad 2 + 2\alpha = \alpha^2 + \alpha^3$$

where α is a root of $1 + \xi^2 = 0$. (Here and in some other places, we use the sign $=$ instead of $\equiv$.)

Some powers of α are equal. The element α is of order 4 and thus generates the cyclic group C_4 rather than the cyclic group C_8. Observe that $p^m - 1 = 8$ is divisible by 4 (see 2.3.5.6). Therefore, α is not a primitive element and $1 + \xi^2$ is not a primitive polynomial (α is not a primitive root of $\xi^8 = 1$).

It can be checked that the element $a = 1+\alpha$ is of order 8 and thus generates the multiplicative group $(\mathbb{GF}(3^2)^*, \times)$ isomorphic to C_8. Therefore, a is a primitive element (a is a primitive root of $\xi^8 = 1$).

$+$	0	a^1	a^2	a^3	a^4	a^5	a^6	a^7	1
0	0	a^1	a^2	a^3	a^4	a^5	a^6	a^7	1
a^1		a^5	1	a^4	a^6	0	a^3	a^2	a^7
a^2			a^6	a^1	a^5	a^7	0	a^4	a^3
a^3				a^7	a^2	a^6	1	0	a^5
a^4					1	a^3	a^7	a^1	0
a^5						a^1	a^4	1	a^2
a^6							a^2	a^5	a^1
a^7								a^3	a^6
1									a^4

Table 2.22. *Addition table for $\mathbb{GF}(3^2) = \mathbb{F}_3[\xi]/\langle 1 + \xi^2 \rangle$ where $a = 1 + \alpha$, $a^2 = 2\alpha$, $a^3 = 1 + 2\alpha$, $a^4 = 2$, $a^5 = 2 + 2\alpha$, $a^6 = \alpha$, $a^7 = 2 + \alpha$ and $a^8 = 1$ (α is a solution of $1 + \xi^2 = 0$); the element at the intersection of the line x and the column y is $x + y$ (the elements below the diagonal of the table are obtained by using $x + y = y + x$)*

The addition and multiplication tables of $\mathbb{GF}(3^2) = \mathbb{F}_3[\xi]/\langle 1 + \xi^2 \rangle$ are given (in terms of powers of $a = 1 + \alpha$) in Tables 2.22 and 2.23, respectively.

– The case $\mathbb{GF}(3^2) = \mathbb{F}_3[\xi]/\langle 2 + \xi + \xi^2 \rangle$. Here, we take

$$P_2(\xi) = 2 + \xi + \xi^2$$

as a prime polynomial in $\mathbb{F}_3[\xi]$. Then, the elements of $\mathbb{GF}(3^2)$ are

$$0, \quad 1 = \beta^8, \quad 2 = \beta^4, \quad \beta, \quad 1 + \beta = \beta^7, \quad 2 + \beta = \beta^6$$

$$2\beta = \beta^5, \quad 1 + 2\beta = \beta^2, \quad 2 + 2\beta = \beta^3$$

where β is a root of $2 + \xi + \xi^2 = 0$.

$\times$	0	a^1	a^2	a^3	a^4	a^5	a^6	a^7	1
0	0	0	0	0	0	0	0	0	0
a^1		a^2	a^3	a^4	a^5	a^6	a^7	1	a^1
a^2			a^4	a^5	a^6	a^7	1	a^1	a^2
a^3				a^6	a^7	1	a^1	a^2	a^3
a^4					1	a^1	a^2	a^3	a^4
a^5						a^2	a^3	a^4	a^5
a^6							a^4	a^5	a^6
a^7								a^6	a^7
1									1

Table 2.23. *Multiplication table for* $\mathbb{GF}(3^2) = \mathbb{F}_3[\xi]/\langle 1 + \xi^2 \rangle$ *where* $a = 1 + \alpha$, $a^2 = 2\alpha$, $a^3 = 1 + 2\alpha$, $a^4 = 2$, $a^5 = 2 + 2\alpha$, $a^6 = \alpha$, $a^7 = 2 + \alpha$ *and* $a^8 = 1$ *(*α *is a solution of* $1 + \xi^2 = 0$*); the element at the intersection of the line* x *and the column* y *is* $x \times y$ *(the elements below the diagonal of the table are obtained by using* $x \times y = y \times x$*)*

All the powers β^s of β are different for $s = 1$ to 8. The element β is of order 8 and generates the group $(\mathbb{GF}(3^2)^*, \times)$. Therefore, β is a primitive element and $2 + \xi + \xi^2$ is a primitive polynomial.

Observe that by introducing $\alpha = 2 + \beta$ into $1 + \alpha^2 = 0$, we obtain $2 + \beta + \beta^2 = 0$ modulo 3. Then, by making the replacements

$$\alpha \to 2 + \beta, \quad a \to \beta$$

in the elements a^1 to $a^8 = 1$ of Tables 2.22 and 2.23, we recover the non-zero elements β, $\beta^2 = 1 + 2\beta$, $\beta^3 = 2 + 2\beta$, $\beta^4 = 2$, $\beta^5 = 2\beta$, $\beta^6 = 2 + \beta$, $\beta^7 = 1 + \beta$ and $\beta^8 = 1$ of $\mathbb{F}_3[\xi]/\langle 2 + \xi + \xi^2 \rangle$.

– The case $\mathbb{GF}(3^2) = \mathbb{F}_3[\xi]/\langle 2 + 2\xi + \xi^2 \rangle$. Here, we take

$$P_2(\xi) = 2 + 2\xi + \xi^2$$

as a prime polynomial in $\mathbb{F}_3[\xi]$. Then, the elements of $\mathbb{GF}(3^2)$ are

$$0, \quad 1 = \gamma^8, \quad 2 = \gamma^4, \quad \gamma, \quad 1 + \gamma = \gamma^2, \quad 2 + \gamma = \gamma^7$$

$$2\gamma = \gamma^5, \quad 1 + 2\gamma = \gamma^3, \quad 2 + 2\gamma = \gamma^6$$

where γ is a root of $2 + 2\xi + \xi^2 = 0$.

All the powers γ^s of γ are different for $s = 1$ to 8. The element γ is of order 8 and generates the group $(\mathbb{GF}(3^2)^*, \times)$. Therefore, γ is a primitive element and $2 + 2\xi + \xi^2$ is a primitive polynomial.

Observe that by introducing $\alpha = 1 + \gamma$ into $1 + \alpha^2 = 0$, we obtain $2 + 2\gamma + \gamma^2 = 0$. Then, by making the replacements

$$\alpha \to 1 + \gamma, \quad a \to \gamma^7$$

in the elements a^1 to $a^8 = 1$ of Tables 2.22 and 2.23, we recover the non-zero elements $\gamma^7 = 2+\gamma$, $\gamma^6 = 2+2\gamma$, $\gamma^5 = 2\gamma$, $\gamma^4 = 2$, $\gamma^3 = 1+2\gamma$, $\gamma^2 = 1+\gamma$, γ and $\gamma^8 = 1$ of $\mathbb{F}_3[\xi]/\langle 2 + 2\xi + \xi^2 \rangle$.

The addition and multiplication tables of the fields $\mathbb{F}_3[\xi]/\langle 2 + \xi + \xi^2 \rangle$ and $\mathbb{F}_3[\xi]/\langle 2 + 2\xi + \xi^2 \rangle$ can be readily built. As a matter of fact, the following correspondences

$$0 \leftrightarrow 0 \leftrightarrow 0, \quad 1 \leftrightarrow 1 \leftrightarrow 1, \quad 1 + \alpha \leftrightarrow \beta \leftrightarrow 2 + \gamma$$

$$2\alpha \leftrightarrow 1 + 2\beta \leftrightarrow 2 + 2\gamma, \quad 1 + 2\alpha \leftrightarrow 2 + 2\beta \leftrightarrow 2\gamma, \quad 2 \leftrightarrow 2 \leftrightarrow 2$$

$$2 + 2\alpha \leftrightarrow 2\beta \leftrightarrow 1 + 2\gamma, \quad \alpha \leftrightarrow 2 + \beta \leftrightarrow 1 + \gamma$$

$$2 + \alpha \leftrightarrow 1 + \beta \leftrightarrow \gamma$$

show that the fields $\mathbb{F}_3[\xi]/\langle 1 + \xi^2 \rangle$, $\mathbb{F}_3[\xi]/\langle 2 + \xi + \xi^2 \rangle$ and $\mathbb{F}_3[\xi]/\langle 2 + 2\xi + \xi^2 \rangle$ are isomorphic. They correspond to the same field $\mathbb{GF}(3^2)$.

2.3.8.8. *Example:* $\mathbb{GF}(3^3) = \mathbb{F}_3[\xi]/\langle 1 + 2\xi + \xi^3 \rangle$

Here we have $p = m = 3$. We can take

$$P_3(\xi) = 1 + 2\xi + \xi^3$$

as prime (i.e. monic + irreducible) polynomial. Then, the $3^3 = 27$ elements of $\mathbb{F}_3[\xi]/\langle 1 + 2\xi + \xi^3 \rangle$ are given by

$$[000] = 0, \quad [001] = \alpha^2, \quad [002] = \alpha^{15}, \quad [010] = \alpha$$
$$[011] = \alpha^{10}, \quad [012] = \alpha^{17}, \quad [020] = \alpha^{14}, \quad [021] = \alpha^4$$
$$[022] = \alpha^{23}, \quad [100] = \alpha^{26}, \quad [101] = \alpha^{21}, \quad [102] = \alpha^{25}$$
$$[110] = \alpha^9, \quad [111] = \alpha^6, \quad [112] = \alpha^{20}, \quad [120] = \alpha^{16}, \; [121] = \alpha^{18}$$
$$[122] = \alpha^{24}, \quad [200] = \alpha^{13}, \quad [201] = \alpha^{12}, \quad [202] = \alpha^8, \; [210] = \alpha^3$$
$$[211] = \alpha^{11}, \quad [212] = \alpha^5, \quad [220] = \alpha^{22}, \quad [221] = \alpha^7, \; [222] = \alpha^{19}$$

where α is a root of $1 + 2\xi + \xi^3 = 0$ and where we use $[x_0 \; x_1 \; x_2]$ to denote the element $x_0 + x_1\alpha + x_2\alpha^2$ of $\mathbb{GF}(3^3)$.

Note that the powers α^s of α for $s = 1$ to 26 constitute the 26 non-zero elements of the field $\mathbb{F}_3[\xi]/\langle 1 + 2\xi + \xi^3 \rangle$. Therefore, α is a primitive element and $1 + 2\xi + \xi^3$ a primitive polynomial.

2.3.8.9. *Example:* $\mathbb{GF}(3^3) = \mathbb{F}_3[\xi]/\langle 2 + 2\xi + \xi^3 \rangle$

Here we have $p = m = 3$. We can take

$$P_3(\xi) = 2 + 2\xi + \xi^3$$

as prime (i.e. monic + irreducible) polynomial. Then, we have

$$\alpha^3 = \alpha^{16} = 1 + \alpha, \; \alpha^4 = \alpha^{17} = \alpha + \alpha^2, \; \alpha^5 = \alpha^{18} = 1 + \alpha + \alpha^2$$
$$\alpha^6 = \alpha^{19} = 1 + 2\alpha + \alpha^2, \; \alpha^7 = \alpha^{20} = 1 + 2\alpha + 2\alpha^2$$
$$\alpha^8 = \alpha^{21} = 2 + 2\alpha^2, \; \alpha^9 = \alpha^{22} = 2 + \alpha$$
$$\alpha^{10} = \alpha^{23} = 2\alpha + \alpha^2, \; \alpha^{11} = \alpha^{24} = 1 + \alpha + 2\alpha^2$$
$$\alpha^{12} = \alpha^{25} = 2 + \alpha^2, \; \alpha^{13} = \alpha^{26} = 1, \; \alpha^{14} = \alpha^{27} = \alpha, \; \alpha^{15} = \alpha^2$$

where α is a root of $2 + 2\xi + \xi^3 = 0$.

Note that some powers α^s $(1 \le s \le 26)$ of α are equal. Therefore, α is not a primitive element and $2 + 2\xi + \xi^3$ is not a primitive polynomial (the element α is of order 13 and 13 is a divisor of $p^m - 1 = 26$).

2.3.8.10. *Example:* $\mathbb{GF}(5^2)$

Here $p = 5$ and $m = 2$. The $5^2 = 25$ elements x of the field $\mathbb{GF}(5^2) = \mathbb{F}_5[\xi]/\langle P_2(\xi)\rangle$ are of the form

$$x = x_0 + x_1\alpha, \quad x_0 \in \mathbb{F}_5,\ x_1 \in \mathbb{F}_5$$

where α is a root of some prime polynomial $P_2(\xi)$ over $\mathbb{F}_5$. We will use the notation $x = [x_0\ x_1]_\alpha$ and examine in turn the cases of the two prime polynomials $P_2(\xi) = 3 + \xi^2$ and $2 + \xi + \xi^2$ of $\mathbb{F}_5[\xi]$.

First, let us take

$$P_2(\xi) = 3 + \xi^2$$

and let us denote the root α of the general case as β. Then, the addition $x + y$ and the multiplication $x \times y$ of the two elements $x = [x_0\ x_1]_\beta$ and $y = [y_0\ y_1]_\beta$ of $\mathbb{F}_5[\xi]/\langle 3 + \xi^2\rangle$ are given by

$$x + y = [x_0 + y_0\ x_1 + y_1]_\beta$$

and

$$x \times y = [x_0 \times y_0 + 2 \times x_1 \times y_1 \quad x_0 \times y_1 + x_1 \times y_0]_\beta$$

The latter two equations show that $\mathbb{F}_5[\xi]/\langle 3 + \xi^2\rangle$ is isomorphic to the field defined in 2.1.2.6 and which corresponds to $\beta = \sqrt{2}$ (a root of $3 + \xi^2 = 0 \bmod 5$). The element β is of order 8 and thus does not generate the cyclic group C_{24} isomorphic to $(\mathbb{GF}(5^2)^*, \times)$. Therefore, β is not a primitive element and $3 + \xi^2$ is not a primitive polynomial.

Second, we take

$$P_2(\xi) = 2 + \xi + \xi^2$$

Observe that $2 + \xi + \xi^2$ can be deduced from $3 + \xi^2$ via the change

$$\xi \to \xi - 2 \bmod 5$$

Let

$$\gamma = 2 + \beta$$

Then, γ is a root of $2 + \xi + \xi^2 = 0$ if β is a root of $3 + \xi^2 = 0$. It can be shown that the element γ is of order 24 and thus generates the group $(\mathbb{GF}(5^2)^*, \times)$. The element γ is thus a primitive element corresponding to the primitive polynomial $2 + \xi + \xi^2$.

For the purpose of comparison, we list in Table 2.24 the non-zero elements of the field $\mathbb{GF}(5^2) = \mathbb{F}_5[\xi]/\langle P_2(\xi)\rangle$ in terms of $[x_0\ x_1]_\beta$ for $P_2(\xi) = 3 + \xi^2$ and $[x_0\ x_1]_\gamma$ for $P_2(\xi) = 2 + \xi + \xi^2$. Table 2.24 shows that β is of order 8 and γ of order 24: $n = 8$ (respectively $n = 24$) is the smallest n such that $\beta^n = 1$ (respectively $\gamma^n = 1$).

$\gamma = [0\ 1]_\gamma = [2\ 1]_\beta$	$\gamma^2 = [3\ 4]_\gamma = [1\ 4]_\beta$	$\gamma^3 = [2\ 4]_\gamma = [0\ 4]_\beta = \beta^5$
$\gamma^4 = [2\ 3]_\gamma = [3\ 3]_\beta$	$\gamma^5 = [4\ 4]_\gamma = [2\ 4]_\beta$	$\gamma^6 = [2\ 0]_\gamma = [2\ 0]_\beta = \beta^2$
$\gamma^7 = [0\ 2]_\gamma = [4\ 2]_\beta$	$\gamma^8 = [1\ 3]_\gamma = [2\ 3]_\beta$	$\gamma^9 = [4\ 3]_\gamma = [0\ 3]_\beta = \beta^7$
$\gamma^{10} = [4\ 1]_\gamma = [1\ 1]_\beta$	$\gamma^{11} = [3\ 3]_\gamma = [4\ 3]_\beta$	$\gamma^{12} = [4\ 0]_\gamma = [4\ 0]_\beta = \beta^4$
$\gamma^{13} = [0\ 4]_\gamma = [3\ 4]_\beta$	$\gamma^{14} = [2\ 1]_\gamma = [4\ 1]_\beta$	$\gamma^{15} = [3\ 1]_\gamma = [0\ 1]_\beta = \beta^1$
$\gamma^{16} = [3\ 2]_\gamma = [2\ 2]_\beta$	$\gamma^{17} = [1\ 1]_\gamma = [3\ 1]_\beta$	$\gamma^{18} = [3\ 0]_\gamma = [3\ 0]_\beta = \beta^6$
$\gamma^{19} = [0\ 3]_\gamma = [1\ 3]_\beta$	$\gamma^{20} = [4\ 2]_\gamma = [3\ 2]_\beta$	$\gamma^{21} = [1\ 2]_\gamma = [0\ 2]_\beta = \beta^3$
$\gamma^{22} = [1\ 4]_\gamma = [4\ 4]_\beta$	$\gamma^{23} = [2\ 2]_\gamma = [1\ 2]_\beta$	$\gamma^{24} = [1\ 0]_\gamma = [1\ 0]_\beta = \beta^8$

Table 2.24. *Non-zero elements of the field $\mathbb{GF}(5^2) = \mathbb{F}_5[\xi]/\langle P_2(\xi)\rangle$ in terms of $[x_0\ x_1]_\beta$ for $P_2(\xi) = 3 + \xi^2$ and $[x_0\ x_1]_\gamma$ for $P_2(\xi) = 2 + \xi + \xi^2$ where β and γ satisfy $3 + \beta^2 = 0$ and $2 + \gamma + \gamma^2 = 0$, respectively*

2.3.9. *Matrix realization of a Galois field*

2.3.9.1. *Linear representation*

PROPOSITION 2.16.– For any Galois field $\mathbb{GF}(p^m) = \mathbb{F}_p[\xi]/\langle P_m(\xi)\rangle$ with $m \geq 2$, it is possible to construct a matrix realization (or linear representation) of the field by matrices of dimension $m \times m$ with matrix elements in $\mathbb{F}_p$. Such a matrix representation is built as follows.

– To the element α, root of $P_m(\xi) = 0$, we associate the $m \times m$ matrix

$$A = \begin{pmatrix} 0 & 0 \cdots 0 & -c_0 \\ 1 & 0 \cdots 0 & -c_1 \\ 0 & 1 \cdots 0 & -c_2 \\ \vdots & \vdots \quad \vdots & \vdots \\ 0 & 0 \cdots 1 & -c_{m-1} \end{pmatrix}$$

where $c_0, c_1, \cdots, c_{m-1}$ are the coefficients in the prime polynomial

$$P_m(\xi) = c_0 + c_1\xi + \cdots + c_{m-1}\xi^{m-1} + \xi^m$$

with $c_0, c_1, \cdots, c_{m-1} \in \mathbb{F}_p$.

– The matrix representation X of the element

$$x = x_0 + x_1\alpha + \cdots + x_{m-1}\alpha^{m-1}$$

of $\mathbb{F}_p[\xi]/\langle P_m(\xi)\rangle$ is then given by

$$X = x_0 I + x_1 A + \cdots + x_{m-1}A^{m-1}$$

where $x_0, x_1, \cdots, x_{m-1} \in \mathbb{F}_p$ and I is the $m \times m$ identity matrix. Through the correspondence $x \leftrightarrow X$, the laws $+$ and $\times$ of $\mathbb{GF}(p^m) = \mathbb{F}_p[\xi]/\langle P_m(\xi)\rangle$ are replaced by the addition and multiplication modulo p of matrices, respectively.

ELEMENT OF PROOF.– The matrix A is the *Frobenius companion matrix* of the polynomial $P_m(\xi)$. Then, it can be checked that the characteristic polynomial of A reads

$$\det(A - \xi I) = (-1)^m P_m(\xi)$$

Therefore, $P_m(A)$, where c_0 is replaced by $c_0 I$, satisfies

$$P_m(A) = O$$

where O is the null $m \times m$ matrix. Then, A constitutes a realization of α and generates via $x \mapsto X$ a representation of $\mathbb{GF}(p^m) = \mathbb{F}_p[\xi]/\langle P_m(\xi)\rangle$. □

The elements of a given field $\mathbb{GF}(p^m)$ can thus be realized either as polynomials $(x_0 + x_1\alpha + \cdots + x_{m-1}\alpha^{m-1})$ or monomials (0 and α^n, $1 \leq n \leq p^m - 1$, if α is a primitive element) or $m \times m$ matrices.

2.3.9.2. *Example: matrix representation of* $\mathbb{GF}(2^3)$

Let us consider the field $\mathbb{GF}(2^3) = \mathbb{F}_2[\xi]/\langle 1 + \xi + \xi^3\rangle$. In this case, the prime polynomial $P_3(\xi)$ of $\mathbb{F}_2[\xi]$ is

$$P_3(\xi) = 1 + \xi + \xi^3 \Rightarrow c_0 = 1, \; c_1 = 1, \; c_2 = 0$$

Therefore, the element α, a root of $1+\xi+\xi^3=0$, is represented by the 3×3 matrix

$$A=\begin{pmatrix}0 & 0 & -1\\ 1 & 0 & -1\\ 0 & 1 & 0\end{pmatrix}\equiv\begin{pmatrix}0 & 0 & 1\\ 1 & 0 & 1\\ 0 & 1 & 0\end{pmatrix} \bmod 2$$

More generally, the element

$$x=x_0+x_1\alpha+x_2\alpha^2$$

of $\mathbb{F}_2[\xi]/\langle 1+\xi+\xi^3\rangle$ is represented by the 3×3 matrix

$$X=x_0I+x_1A+x_2A^2$$

This yields the following representation

$$0\leftrightarrow O=\begin{pmatrix}0 & 0 & 0\\ 0 & 0 & 0\\ 0 & 0 & 0\end{pmatrix},\quad 1\leftrightarrow I=\begin{pmatrix}1 & 0 & 0\\ 0 & 1 & 0\\ 0 & 0 & 1\end{pmatrix}$$

$$a\leftrightarrow A=\begin{pmatrix}0 & 0 & 1\\ 1 & 0 & 1\\ 0 & 1 & 0\end{pmatrix},\quad b\leftrightarrow B=A^2=\begin{pmatrix}0 & 1 & 0\\ 0 & 1 & 1\\ 1 & 0 & 1\end{pmatrix}$$

$$c\leftrightarrow C=I+A=\begin{pmatrix}1 & 0 & 1\\ 1 & 1 & 1\\ 0 & 1 & 1\end{pmatrix},\quad d\leftrightarrow D=A+A^2=\begin{pmatrix}0 & 1 & 1\\ 1 & 1 & 0\\ 1 & 1 & 1\end{pmatrix}$$

$$e\leftrightarrow E=I+A+A^2=\begin{pmatrix}1 & 1 & 1\\ 1 & 0 & 0\\ 1 & 1 & 0\end{pmatrix}$$

$$f\leftrightarrow F=I+A^2=\begin{pmatrix}1 & 1 & 0\\ 0 & 0 & 1\\ 1 & 0 & 0\end{pmatrix}$$

of the field $\mathbb{F}_2[\xi]/\langle 1+\xi+\xi^3\rangle$ (the elements $a,b,\cdots,f$ are defined in Example 2.3.8.4). As a check, it can be seen that the addition and multiplication tables modulo 2 of the matrices $O,I,A,\cdots,F$ are identical to Tables 2.17 and 2.18, respectively (with the replacements $0 \to O$, $1 \to I$, $a \to A$, $\cdots$, $f \to F$). Note that

$$a \leftrightarrow A, \quad b \leftrightarrow B = A^2, \quad c \leftrightarrow C = A^3$$

$$d \leftrightarrow D = A^4, \quad e \leftrightarrow E = A^5, \quad f \leftrightarrow F = A^6, \quad 1 \leftrightarrow I = A^7$$

Therefore, the matrix A generates a group (with respect to matrix multiplication) isomorphic to the cyclic group C_7, in agreement with the fact that α is a primitive element.

2.3.9.3. *Example: matrix representation of* $\mathbb{GF}(3^2)$

First, let us consider the realization $\mathbb{F}_3[\xi]/\langle 1 + \xi^2 \rangle$ of the field $\mathbb{GF}(3^2)$. From the prime polynomial

$$P_2(\xi) = 1 + \xi^2 \Rightarrow c_0 = 1, \ c_1 = 0$$

of $\mathbb{F}_3[\xi]$, it follows that the element α, a root of $1 + \xi^2 = 0$, is represented by the 2×2 matrix A:

$$\alpha \leftrightarrow A = \begin{pmatrix} 0 & -1 \\ 1 & 0 \end{pmatrix} \equiv \begin{pmatrix} 0 & 2 \\ 1 & 0 \end{pmatrix} \mod 3$$

The element x of $\mathbb{F}_3[\xi]/\langle 1 + \xi^2 \rangle$ is thus represented by the 2×2 matrix X:

$$x = x_0 + x_1\alpha \leftrightarrow X = x_0 I + x_1 A$$

This yields the matrix representation

$$0 \leftrightarrow O = \begin{pmatrix} 0 & 0 \\ 0 & 0 \end{pmatrix}, \quad 1 \leftrightarrow I = \begin{pmatrix} 1 & 0 \\ 0 & 1 \end{pmatrix}, \quad 2 \leftrightarrow 2I = \begin{pmatrix} 2 & 0 \\ 0 & 2 \end{pmatrix}$$

$$\alpha \leftrightarrow A = \begin{pmatrix} 0 & 2 \\ 1 & 0 \end{pmatrix}, \quad 1 + \alpha \leftrightarrow I + A = \begin{pmatrix} 1 & 2 \\ 1 & 1 \end{pmatrix}$$

$$2 + \alpha \leftrightarrow 2I + A = \begin{pmatrix} 2 & 2 \\ 1 & 2 \end{pmatrix}, \quad 2\alpha \leftrightarrow 2A = \begin{pmatrix} 0 & 1 \\ 2 & 0 \end{pmatrix}$$

$$1 + 2\alpha \leftrightarrow I + 2A = \begin{pmatrix} 1 & 1 \\ 2 & 1 \end{pmatrix}, \quad 2 + 2\alpha \leftrightarrow 2I + 2A = \begin{pmatrix} 2 & 1 \\ 2 & 2 \end{pmatrix}$$

of the field $\mathbb{F}_3[\xi]/\langle 1+\xi^2\rangle$. The matrix A generates (with respect to matrix multiplication) the cyclic group C_4 ($A^4 = I \Rightarrow A^5 = A$, $A^6 = A^2$, $A^7 = A^3$, $A^8 = A^4 = I$, note α is not a primitive element).

As the second realization, let us take $\mathbb{GF}(3^2) = \mathbb{F}_3[\xi]/\langle 2+2\xi+\xi^2\rangle$. Then

$$P_2(\xi) = 2 + 2\xi + \xi^2 \Rightarrow c_0 = 2, \ c_1 = 2$$

so that

$$\alpha \leftrightarrow A = \begin{pmatrix} 0 & -2 \\ 1 & -2 \end{pmatrix} \equiv \begin{pmatrix} 0 & 1 \\ 1 & 1 \end{pmatrix} \bmod 3$$

where α is a root of $2 + 2\xi + \xi^2 = 0$. This leads to the matrix representation

$$0 \leftrightarrow O = \begin{pmatrix} 0 & 0 \\ 0 & 0 \end{pmatrix}, \quad 1 \leftrightarrow I = \begin{pmatrix} 1 & 0 \\ 0 & 1 \end{pmatrix}, \quad 2 \leftrightarrow 2I = \begin{pmatrix} 2 & 0 \\ 0 & 2 \end{pmatrix}$$

$$\alpha \leftrightarrow A = \begin{pmatrix} 0 & 1 \\ 1 & 1 \end{pmatrix}, \quad 1 + \alpha \leftrightarrow I + A = \begin{pmatrix} 1 & 1 \\ 1 & 2 \end{pmatrix}$$

$$2 + \alpha \leftrightarrow 2I + A = \begin{pmatrix} 2 & 1 \\ 1 & 0 \end{pmatrix}, \quad 2\alpha \leftrightarrow 2A = \begin{pmatrix} 0 & 2 \\ 2 & 2 \end{pmatrix}$$

$$1 + 2\alpha \leftrightarrow I + 2A = \begin{pmatrix} 1 & 2 \\ 2 & 0 \end{pmatrix}, \quad 2 + 2\alpha \leftrightarrow 2I + 2A = \begin{pmatrix} 2 & 2 \\ 2 & 1 \end{pmatrix}$$

of the field $\mathbb{F}_3[\xi]/\langle 2+2\xi+\xi^2\rangle$. Here, we have

$$A, \quad I + A = A^2, \quad I + 2A = A^3, \quad 2I = A^4$$

$$2A = A^5, \quad 2I + 2A = A^6, \quad 2I + A = A^7, \quad I = A^8$$

and thus the matrix A generates (with respect to matrix multiplication) the cyclic group C_8 (see α is a primitive element).

2.4. Sub-field of a Galois field

2.4.1. $\mathbb{GF}(p^\ell)$ *sub-field of* $\mathbb{GF}(p^m)$

2.4.1.1. *Necessary and sufficient condition for* $\mathbb{GF}(p^\ell) \subset \mathbb{GF}(p^m)$

PROPOSITION 2.17.– Any Galois field $\mathbb{GF}(p^m)$, of cardinal p^m, contains the field $\mathbb{F}_p$ (isomorphic to $\mathbb{Z}_p$) as a sub-field. Indeed, the prime field $\mathbb{F}_p$ is the smallest sub-field of $\mathbb{GF}(p^m)$. In terms of cardinals, we have that

$$|\mathbb{GF}(p^m)| = |\mathbb{F}_p|^m$$

(we use $|\mathbb{K}|$ to denote the cardinal of the field $\mathbb{K}$). More generally, $\mathbb{GF}(p^\ell)$ is a sub-field of $\mathbb{GF}(p^m)$, $\mathbb{GF}(p^\ell) \subset \mathbb{GF}(p^m)$, if and only if ℓ is a divisor of m. The case $\ell = 1$ corresponds to $\mathbb{F}_p \subset \mathbb{GF}(p^m)$.

PROOF.– The main part of the proposition rests on the fact that $p^\ell - 1$, the order of the group $(\mathbb{GF}(p^\ell)^*, \times)$, divides $p^m - 1$, the order of the group $(\mathbb{GF}(p^m)^*, \times)$, if and only if ℓ divides m. □

It is important to note that the number of sub-fields of $\mathbb{GF}(p^m)$ is equal to the number of positive divisors of m.

2.4.1.2. *Example: sub-field of* $\mathbb{F}_p$

The sole sub-field of $\mathbb{F}_p$ is $\mathbb{F}_p$ (the field $\mathbb{F}_p$ has no proper sub-field).

2.4.1.3. *Example: sub-fields of* $\mathbb{GF}(2^4)$

For the field $\mathbb{GF}(2^4) = \mathbb{F}_2[\xi]/\langle 1 + \xi + \xi^4 \rangle$, we can check that the set

$$\{0, 1, \alpha^5 = \alpha + \alpha^2, \alpha^{10} = 1 + \alpha + \alpha^2\}$$

endowed with the laws $+$ and $\times$ of $\mathbb{GF}(2^4)$ is a sub-field of $\mathbb{GF}(2^4)$ that is isomorphic to $\mathbb{GF}(2^2)$, in agreement with the fact that 2 divides 4. The field $\mathbb{GF}(2^4)$ admits three sub-fields, namely, $\mathbb{F}_2$, $\mathbb{GF}(2^2)$ and $\mathbb{GF}(2^4)$ yielding the chain of fields $\mathbb{F}_2 \subset \mathbb{GF}(2^2) \subset \mathbb{GF}(2^4)$.

2.4.1.4. *Example: sub-fields of* $\mathbb{GF}(2^6)$

The sub-fields of $\mathbb{GF}(2^6)$ are of type $\mathbb{GF}(2^\ell)$ where ℓ divides 6. This gives four sub-fields, viz. $\mathbb{F}_2$, $\mathbb{GF}(2^2)$, $\mathbb{GF}(2^3)$ and $\mathbb{GF}(2^6)$. We thus have the chains of fields $\mathbb{F}_2 \subset \mathbb{GF}(2^2) \subset \mathbb{GF}(2^6)$ and $\mathbb{F}_2 \subset \mathbb{GF}(2^3) \subset \mathbb{GF}(2^6)$.

2.4.2. *Characteristic of the sub-fields of* $\mathbb{GF}(p^m)$

PROPOSITION 2.18.– Every sub-field of $\mathbb{GF}(p^m)$, the characteristic of which is p, has the characteristic p.

Remember: for all x in the field $\mathbb{GF}(p^m)$ with $m \geq 1$, we have

$$x + x + \cdots + x = px \equiv 0 \bmod p$$

where the sum contains p summands x. This equality is also true for any element belonging to a sub-field of $\mathbb{GF}(p^m)$.

Propositions 2.4.1.1 and 2.4.2 show that all the sub-fields of $\mathbb{GF}(p^m)$ are of the type $\mathbb{GF}(p^\ell)$ where ℓ divides m.

2.5. Factorizations

We give below some propositions valid for the field $\mathbb{GF}(p^m)$ with p prime (even or odd) and m positive integer ($m \geq 1$). They can be seen to be satisfied by the examples given in 1.2.5.

2.5.1. *Powers of elements of* $\mathbb{GF}(p^m)$

2.5.1.1. *Remarkable identities*

PROPOSITION 2.19.– In the field $\mathbb{GF}(p^m)$, we have

$$\forall x \in \mathbb{GF}(p^m)^* : x^{p^m - 1} \equiv 1, \quad \forall x \in \mathbb{GF}(p^m) : x^{p^m} \equiv x$$

$$\forall x \in \mathbb{GF}(p^m),\ \forall y \in \mathbb{GF}(p^m) : (x \pm y)^{p^n} \equiv x^{p^n} \pm y^{p^n}$$

$$(xy)^{p^n} = x^{p^n} y^{p^n}$$

for n in $\mathbb{N}_0$ (all identities indicated by $\equiv$ are valid modulo p). In the special case $n = m$, we have

$$(x \pm y)^{p^m} \equiv x^{p^m} \pm y^{p^m} \equiv x \pm y, \quad (xy)^{p^m} = x^{p^m} y^{p^m} \equiv xy$$

In particular,

$$\forall a \in \mathbb{F}_p,\ \forall b \in \mathbb{F}_p : a^p \equiv a,\ (a \pm b)^p \equiv a^p \pm b^p \equiv a \pm b,$$

$$(ab)^p = a^p b^p \equiv ab$$

for $m = 1$.

PROOF.– Lagrange's theorem for finite groups implies $x^{p^m-1} \equiv 1$ for all x in the group $(\mathbb{GF}(p^m)^*, \times)$ of order $p^m - 1$. Therefore, $x^{p^m} \equiv x$ for all $x \in \mathbb{GF}(p^m)$. As a particular case, for $m = 1$, the congruence relation $x^{p^m} \equiv x$ for all $x \in \mathbb{GF}(p^m)$ gives $a^p \equiv a$ for all $a \in \mathbb{GF}(p) = \mathbb{F}_p$. For $x \in \mathbb{GF}(p^m)$ and $y \in \mathbb{GF}(p^m)$, the binomial theorem yields $(x+y)^p \equiv x^p + y^p \bmod p$ since, for p prime, p is a divisor of the binomial coefficients C_p^k for $1 \leq k \leq p-1$; the identity $(x+y)^{p^n} \equiv x^{p^n} + y^{p^n}$ with $2 \leq n \leq m$ can be proved by induction for $m \geq 2$. The rest of the proof is trivial. □

2.5.1.2. *Remarks*

– For all x in $\mathbb{GF}(p^m)$ with $m \geq 2$, the relation $x^{p^m} \equiv x$ holds, but $x^p \neq x$ in general. An element x of $\mathbb{GF}(p^m)$ belongs to $\mathbb{F}_p$ if and only if $x^p = x$.

– For $m = 1$, the relation $a^p \equiv a$ for all $a \in \mathbb{F}_p$ is nothing but the Fermat little theorem according to which

$$p \text{ divides } a^p - a \Leftrightarrow a^p \equiv a \bmod p$$

for any integer a and any prime p and

$$a^{p-1} \equiv 1 \bmod p$$

for any integer a not divisible by p prime ($p \not| \, a$).

2.5.2. *Solutions of* $\xi^{p^m} - \xi = 0$

2.5.2.1. *Roots of* $\xi^{p^m-1} - 1$

PROPOSITION 2.20.– The non-zero elements of $\mathbb{GF}(p^m)$ are the $p^m - 1$ distinct roots of the polynomial $\xi^{p^m-1} - 1$ in $\mathbb{GF}(p^m)[\xi]$. That is to say, every non-zero element x of the Galois field $\mathbb{GF}(p^m)$ satisfies

$$\xi^{p^m-1} \equiv 1 \bmod p$$

and is thus a $p^m - 1$-th root of unity. Therefore,

$$\xi^{p^m-1} - 1 \equiv \prod_{x \in \mathbb{GF}(p^m)^*} (\xi - x) \bmod p$$

where the sum does not involve the element $x = 0$.

ELEMENT OF PROOF.– Note that the polynomial $\xi^{p^m-1} - 1$ does not have repeated root in $\mathbb{GF}(p^m)[\xi]$ since the derivative

$$\left(\xi^{p^m-1} - 1\right)' = (p^m - 1)\xi^{p^m-2}$$

vanishes for $\xi = 0$ only and $\xi = 0$ is not a root of $\xi^{p^m-1} - 1$. (Remember that in ordinary algebra, a one-variable polynomial has a double root if and only if the polynomial and its derivative have a common root.) Proposition 2.5.2.1 reflects the fact that the non-zero elements of $\mathbb{GF}(p^m)$ form a cyclic group of order $p^m - 1$. □

As an immediate corollary, we have the following result.

2.5.2.2. *Roots of* $\xi^{p^m} - \xi$

PROPOSITION 2.21.– Each of the p^m elements of the field $\mathbb{GF}(p^m)$ is a root of the equation

$$\xi^{p^m} - \xi \equiv 0 \bmod p$$

Therefore, $\xi^{p^m} - \xi$ can be factored into linear factors (monic polynomials of degree 1) in $\mathbb{GF}(p^m)[\xi]$ as

$$\xi^{p^m} - \xi \equiv \prod_{x \in \mathbb{GF}(p^m)} (\xi - x) \bmod p$$

In a more detailed way,

$$\xi^{p^m} - \xi \equiv \prod_{i=1}^{p^m} (\xi - x_i) = (\xi - x_1)(\xi - x_2) \cdots (\xi - x_{p^m}) \bmod p$$

where $x_1, x_2, \cdots, x_{p^m}$ are the p^m elements of $\mathbb{GF}(p^m)$; alternatively,

$$\xi^{p^m} - \xi \equiv \xi(\xi - 1)(\xi - \alpha)(\xi - \alpha^2) \cdots (\xi - \alpha^{p^m-2}) \bmod p$$

where α is a primitive element of $\mathbb{GF}(p^m)$.

PROOF.– First, note that $\xi = 0$ and $\xi = 1$ are elementary solutions of the equation $\xi^{p^m} - \xi = 0$. Second, it is easy to recover that the other elements of $\mathbb{GF}(p^m)$ are solutions of $\xi^{p^m} - \xi = 0$. In this respect, let α be a primitive element of the field $\mathbb{GF}(p^m)$; then, for any element α^s with $s = 1, 2, \cdots, p^m - 2$, we have

$$(\alpha^s)^{p^m} - \alpha^s = (\alpha^{p^m})^s - \alpha^s \equiv \alpha^s - \alpha^s = 0 \bmod p$$

Therefore, every element of $\mathbb{GF}(p^m)$ is a solution of $\xi^{p^m} - \xi = 0$. Equivalently, the decomposition

$$\xi^{p^m} - \xi \equiv \prod_{x \in \mathbb{GF}(p^m)} (\xi - x) \bmod p$$

holds in the ring of polynomials $\mathbb{GF}(p^m)[\xi]$. Thus, $\xi^{p^m} - \xi$ has p^m distinct roots (viz. the root $\xi = 0$ plus the $p^m - 1$ solutions of $\xi^{p^m - 1} - 1 = 0$). □

It is straightforward to check that if x_1 and x_2 are roots of the equation $\xi^{p^m} - \xi \equiv 0 \bmod p$, then $x_1 + x_2$ and $x_1 x_2$ are roots equally well.

2.5.2.3. *Example:* $\mathbb{F}_p$

For the field $\mathbb{F}_p$, the relation

$$\xi^p - \xi \equiv \prod_{a \in \mathbb{F}_p} (\xi - a) \bmod p$$

follows from Proposition 2.5.2.2 with $m = 1$. For instance,

$$(\xi - 0)(\xi - 1)(\xi - 2)(\xi - 3)(\xi - 4) \equiv \xi^5 - \xi \bmod 5$$

in the case of $\mathbb{F}_5$.

2.5.2.4. *Example:* $\mathbb{GF}(2^2) = \mathbb{F}_2[\xi]/\langle 1 + \xi + \xi^2 \rangle$

In the case $p = 2$ and $m = 2$, we can check that

$$(\xi - 0)(\xi - 1)(\xi - \alpha)(\xi - 1 - \alpha) \equiv \xi^4 - \xi \bmod 2$$

or

$$(\xi - 0)(\xi - 1)(\xi - \alpha)(\xi - \alpha^2) \equiv \xi^4 - \xi \bmod 2$$

where the primitive element α is a root of $1 + \xi + \xi^2 = 0$.

2.5.2.5. *Example:* $\mathbb{GF}(2^3) = \mathbb{F}_2[\xi]/\langle 1+\xi+\xi^3\rangle$

In the case $p = 2$ and $m = 3$, we verify that

$$(\xi-0)(\xi-1)(\xi-\alpha)(\xi-\alpha^2)(\xi-1-\alpha)(\xi-\alpha-\alpha^2)$$
$$\times(\xi-1-\alpha-\alpha^2)(\xi-1-\alpha^2) \equiv \xi^8-\xi \bmod 2$$

or

$$(\xi-0)(\xi-1)(\xi-\alpha)(\xi-\alpha^2)(\xi-\alpha^3)(\xi-\alpha^4)$$
$$\times(\xi-\alpha^5)(\xi-\alpha^6) \equiv \xi^8-\xi \bmod 2$$

where the primitive element α is a root of $1+\xi+\xi^3 = 0$.

2.5.2.6. *Example:* $\mathbb{GF}(3^2) = \mathbb{F}_3[\xi]/\langle 1+\xi^2\rangle$

In the case $p = 3$ and $m = 2$, we verify that

$$(\xi-0)(\xi-1)(\xi-2)(\xi-\alpha)(\xi-1-\alpha)(\xi-2-\alpha)(\xi-2\alpha)$$
$$\times(\xi-1-2\alpha)(\xi-2-2\alpha) \equiv \xi^9-\xi \bmod 3$$

where the element α is a root of $1+\xi^2 = 0$ (α is not a primitive element).

2.5.3. *Product of all the elements of* $\mathbb{GF}(p^m)^*$

2.5.3.1. *Wilson's theorem*

PROPOSITION 2.22.– The product of all the elements of $\mathbb{GF}(p^m)^*$ with $m \geq 1$ is equal to -1 modulo p.

PROOF.– The non-zero elements of $\mathbb{GF}(p^m)$ are the roots of the equation $\xi^{p^m-1} - 1 = 0$ for which the product of the $p^m - 1$ roots is

– $+1 \equiv -1 \bmod 2$ for p even ($p = 2$);

– -1 for p odd ($p = 3, 5, 7, \cdots$).

Observe that Proposition 2.5.3.1 holds for any even or odd prime p. □

As an immediate corollary, Proposition 2.5.3.1 applied to the field $\mathbb{F}_p$ yields

$$(p-1)! \equiv -1 \bmod p$$

for any even or odd prime p, a result known as Wilson's theorem.

2.5.3.2. *Examples*

The product of the non-zero elements of $\mathbb{GF}(2^2) = \mathbb{F}_2[\xi]/\langle 1+\xi+\xi^2 \rangle$ is

$$1 \times \alpha \times (1+\alpha) = \alpha + \alpha^2 = -1$$

and the product of the non-zero elements of $\mathbb{GF}(2^3) = \mathbb{F}_2[\xi]/\langle 1+\xi+\xi^3 \rangle$ is

$$1 \times \alpha \times \alpha^2 \times \alpha^3 \times \alpha^4 \times \alpha^5 \times \alpha^6 = \alpha^{21} = 1 \equiv -1 \bmod 2$$

where, in both examples, α is a primitive element (α is a root of $1+\xi+\xi^2 = 0$ for $\mathbb{GF}(2^2)$ and a root of $1+\xi+\xi^3 = 0$ for $\mathbb{GF}(2^3)$). Similarly, the product of the non-zero elements of the field $\mathbb{GF}(3^2) = \mathbb{F}_3[\xi]/\langle 1+\xi^2 \rangle$ can be calculated to be

$$\begin{aligned} &1 \times 2 \times \alpha \times (1+\alpha) \times (2+\alpha) \times 2\alpha \times (1+2\alpha) \times (2+2\alpha) \\ &\equiv -1 \bmod 3 \end{aligned}$$

where α is a root of $1+\xi^2 = 0$ (α is not a primitive element).

2.5.4. *Factorization of $\xi^{p^m} - \xi$ in prime polynomials*

2.5.4.1. *Prime polynomials and $\xi^{p^m} - \xi$*

PROPOSITION 2.23.– For every Galois field $\mathbb{GF}(p^m)$, the polynomial $\xi^{p^m} - \xi$ can be factored as

$$\xi^{p^m} - \xi \equiv \prod_{k=1,\ k|m}^{m} P_k(\xi) \bmod p$$

where the second member is the product of the distinct prime polynomials $P_k(\xi)$ in $\mathbb{F}_p[\xi]$ whose degree k divides m (with no repeated factor). Therefore, if k is a divisor of m, the prime polynomial $P_k(\xi)$ divides $\xi^{p^m} - \xi$. In particular, the prime polynomial $P_m(\xi)$ of degree m is a divisor of $\xi^{p^m} - \xi$.

It is important to note that only the values of k that divide m (including $k = 1$ and $k = m$) occur in the decomposition of $\xi^{p^m} - \xi$ as can be verified in the following examples.

Note also that each element of $\mathbb{GF}(p^m)$ is a root of one polynomial occurring in the development of $\xi^{p^m} - \xi$ in terms of prime polynomials $P_k(\xi)$.

The proposition provides us with a test to know if a monic polynomial of degree m in $\mathbb{F}_p[\xi]$ is irreducible or not. Indeed, such a polynomial is prime (and therefore irreducible) if and only if it divides $\xi^{p^m} - \xi$. Furthermore, if a monic polynomial in $\mathbb{F}_p[\xi]$ of degree k, k being a divisor of m, does not divide $\xi^{p^m} - \xi$, then this polynomial is reducible.

The prime polynomials $P_k(\xi)$ occurring in the factorization of $\xi^{p^m} - \xi$ are called *minimal polynomials* of $\mathbb{GF}(p^m)$.

2.5.4.2. *Example:* $\mathbb{F}_2 = \mathbb{GF}(2^1)$

In the case $p = 2$ and $m = 1$, we have

$$\xi^2 - \xi \equiv \xi(1 + \xi) \bmod 2$$

so that $\xi^2 - \xi$ is the product of the two prime polynomials of $\mathbb{F}_2[\xi]$ of degree 1. These two polynomials are the minimal polynomials of $\mathbb{F}_2$.

2.5.4.3. *Example:* $\mathbb{GF}(2^2)$

In the case $p = 2$ and $m = 2$, we have

$$\xi^4 - \xi \equiv \xi(1 + \xi)(1 + \xi + \xi^2) \bmod 2$$

so that $\xi^4 - \xi$ is the product of the three prime polynomials of $\mathbb{F}_2[\xi]$ of degree lower than or equal to 2. These three polynomials are the minimal polynomials of $\mathbb{GF}(2^2)$.

2.5.4.4. *Example:* $\mathbb{GF}(2^3)$

In the case $p = 2$ and $m = 3$, we have

$$\xi^8 - \xi \equiv \xi(1 + \xi)(1 + \xi + \xi^3)(1 + \xi^2 + \xi^3) \bmod 2$$

so that $\xi^8 - \xi$ is the product of four of the five prime polynomials of $\mathbb{F}_2[\xi]$ of degree lower than or equal to 3. Note that the prime polynomial $1+\xi+\xi^2$ does not appear in the factorization of $\xi^8 - \xi$ because $2 \nmid 3$. The four polynomials occurring in the factorization of $\xi^8-\xi$ are the minimal polynomials of $\mathbb{GF}(2^3)$.

2.5.4.5. *Example:* $\mathbb{GF}(2^4)$

In the case $p = 2$ and $m = 4$, we have

$$\begin{aligned}\xi^{16} - \xi \equiv{}& \xi(1+\xi)(1+\xi+\xi^2)(1+\xi+\xi^4)(1+\xi^3+\xi^4)\\ &\times(1+\xi+\xi^2+\xi^3+\xi^4) \bmod 2\end{aligned}$$

so that $\xi^{16} - \xi$ is the product of six of the eight prime polynomials of $\mathbb{F}_2[\xi]$ of degree lower than or equal to 4. Note that the prime polynomials $1+\xi+\xi^3$ and $1+\xi^2+\xi^3$ do not appear in the factorization of $\xi^{16} - \xi$ because $3 \nmid 4$. The six polynomials occurring in the factorization of $\xi^{16} - \xi$ are the minimal polynomials of $\mathbb{GF}(2^4)$.

2.5.4.6. *Example:* $\mathbb{GF}(2^5)$

In the case $p = 2$ and $m = 5$, we have

$$\begin{aligned}\xi^{32} - \xi \equiv{}& \xi(1+\xi)(1+\xi^2+\xi^5)(1+\xi^3+\xi^5)\\ &\times(1+\xi+\xi^2+\xi^3+\xi^5)(1+\xi+\xi^2+\xi^4+\xi^5)\\ &\times(1+\xi+\xi^3+\xi^4+\xi^5)(1+\xi^2+\xi^3+\xi^4+\xi^5) \bmod 2\end{aligned}$$

so that $\xi^{32} - \xi$ is the product of eight prime polynomials of $\mathbb{F}_2[\xi]$ (the two prime polynomials of degree 1 and the six prime polynomials of degree 5).

2.5.4.7. *Example:* $\mathbb{F}_3 = \mathbb{GF}(3^1)$

In the case $p = 3$ and $m = 1$, we have

$$\xi^3 - \xi \equiv \xi(1+\xi)(2+\xi) \bmod 3$$

so that $\xi^3 - \xi$ is the product of the three prime polynomials of $\mathbb{F}_3[\xi]$ of degree 1.

2.5.4.8. *Example:* $\mathbb{GF}(3^2)$

In the case $p = 3$ and $m = 2$, we have

$$\xi^9 - \xi \equiv \xi(1+\xi)(2+\xi)(1+\xi^2)(2+\xi+\xi^2)(2+2\xi+\xi^2) \bmod 3$$

so that $\xi^9 - \xi$ is the product of the six prime polynomials of $\mathbb{F}_3[\xi]$ of degree lower than or equal to 2.

2.5.4.9. *Example:* $\mathbb{GF}(3^3)$

In the case $p = 3$ and $m = 3$, we have

$$\begin{aligned} x^{27} - x = {} & \xi(1+\xi)(2+\xi)(1+2\xi+\xi^3)(2+2\xi+\xi^3) \\ & \times(2+\xi+\xi^2+\xi^3)(1+2\xi^2+\xi^3)(2+\xi^2+\xi^3) \\ & \times(1+\xi+2\xi^2+\xi^3)(1+2\xi+\xi^2+\xi^3) \\ & \times(2+2\xi+2\xi^2+\xi^3) \bmod 3 \end{aligned}$$

so that $\xi^{27} - \xi$ can be written as a product involving the three prime polynomials of degree 1 and the eight prime polynomials of degree 3 in $\mathbb{F}_3[\xi]$.

2.5.5. *Factorization of a prime polynomial*

2.5.5.1. *Roots of a prime polynomial*

PROPOSITION 2.24.– Each prime polynomial $P_m(\xi)$ of degree m in $\mathbb{F}_p[\xi]$ can be factored as

$$P_m(\xi) \equiv \prod_{j=0}^{m-1} (\xi - \alpha^{p^j}) = (\xi - \alpha)(\xi - \alpha^p) \cdots (\xi - \alpha^{p^{m-1}})$$

modulo p and taking into account $P_m(\alpha) = 0$ where α is a root of $P_m(\xi)$. Therefore, if α is a root of the prime polynomial $P_m(\xi)$, then $\alpha^{p^1}, \alpha^{p^2}, \cdots, \alpha^{p^{m-1}}$ are also roots of $P_m(\xi)$ so that the m distinct elements

$$\alpha^{p^0} = \alpha, \quad \alpha^{p^1} = \alpha^p, \quad \alpha^{p^2}, \quad \cdots, \quad \alpha^{p^{m-1}}$$

of $\mathbb{GF}(p^m) = \mathbb{F}_p[\xi]/\langle P_m(\xi)\rangle$ are the m simple roots of $P_m(\xi)$.

PROOF.– Let α be a root of the prime polynomial $P_m(\xi)$ of degree m in $\mathbb{F}_p[\xi]$

$$P_m(\xi) = c_0 + c_1\xi + c_2\xi^2 + \cdots + c_{m-1}\xi^{m-1} + \xi^m$$

Therefore, $P_m(\alpha) = 0$ and

$$\begin{aligned} P_m(\alpha^p) &= c_0 + c_1\alpha^p + c_2\alpha^{2p} + \cdots + c_{m-1}\alpha^{(m-1)p} + \alpha^{mp} \\ &= c_0{}^p + c_1{}^p\alpha^p + c_2{}^p\alpha^{2p} + \cdots + c_{m-1}{}^p\alpha^{(m-1)p} + \alpha^{mp} \\ &= (c_0 + c_1\alpha + c_2\alpha^2 + \cdots + c_{m-1}\alpha^{(m-1)} + \alpha^m)^p \\ &= (P_m(\alpha))^p \end{aligned}$$

so that $P_m(\alpha^p) = 0$. Then, α^p is a root of $P_m(\xi)$ too. It follows that $(\alpha^p)^p = \alpha^{p^2}$, $(\alpha^{p^2})^p = \alpha^{p^3}, \cdots, (\alpha^{p^{m-2}})^p = \alpha^{p^{m-1}}$ are also roots of $P_m(\xi)$. □

In the case where α, root of a polynomial $P_m(\xi)$ in $\mathbb{F}_p[\xi]$, is a primitive element of $\mathbb{GF}(p^m)$, the m roots $\alpha, \alpha^p, \alpha^{p^2}, \cdots, \alpha^{p^{m-1}}$ of $P_m(\xi)$ are primitive elements of $\mathbb{GF}(p^m)$.

2.5.5.2. *Example:* $\mathbb{GF}(2^2)$

For $\mathbb{GF}(2^2) = \mathbb{F}_2[\xi]/\langle 1 + \xi + \xi^2\rangle$, i.e. $p = 2$ and $m = 2$, we have

$$(\xi - \alpha)(\xi - \alpha^2) = \xi^2 - (\alpha + \alpha^2)\xi + \alpha^3$$

By using

$$1 + \alpha + \alpha^2 = 0 \Rightarrow \alpha + \alpha^2 = -1$$

and

$$\alpha(1 + \alpha + \alpha^2) = 0 \Rightarrow \alpha^3 = -\alpha - \alpha^2 = 1$$

we obtain

$$(\xi - \alpha)(\xi - \alpha^2) = 1 + \xi + \xi^2$$

which is the prime polynomial $P_2(\xi)$ for $\mathbb{GF}(2^2)$.

2.5.5.3. *Example:* $\mathbb{GF}(3^2)$

For $\mathbb{GF}(3^2)$, i.e. $p = 3$ and $m = 2$, we have

$$(\xi - \alpha)(\xi - \alpha^3) = \xi^2 - (\alpha + \alpha^3)\xi + \alpha^4$$

Three choices need to be considered.

– Choice $P_2(\xi) = 1 + \xi^2$. Here $1 + \alpha^2 = 0$. Thus,

$$\alpha(1 + \alpha^2) = 0 \Rightarrow \alpha + \alpha^3 = 0, \quad \alpha^2 = -1 \Rightarrow \alpha^4 = 1$$

Consequently, $\xi^2 - (\alpha + \alpha^3)\xi + \alpha^4 = 1 + \xi^2$ and $P_2(\xi) = (\xi - \alpha)(\xi - \alpha^3)$ is satisfied.

– Choice $P_2(\xi) = 2 + \xi + \xi^2$. Here $2 + \alpha + \alpha^2 = 0$. Thus,

$$\alpha(2 + \alpha + \alpha^2) = 0 \Rightarrow \alpha + \alpha^3 = -\alpha - \alpha^2 = 2 \equiv -1 \bmod 3$$

$$\alpha + \alpha^3 \equiv -1 \Rightarrow \alpha(\alpha + \alpha^3) \equiv -\alpha \Rightarrow \alpha^4 \equiv -\alpha - \alpha^2 = 2 \bmod 3$$

Consequently, $\xi^2 - (\alpha + \alpha^3)\xi + \alpha^4 = 2 + \xi + \xi^2$ and $P_2(\xi) = (\xi - \alpha)(\xi - \alpha^3)$ is satisfied.

– Choice $P_2(\xi) = 2 + 2\xi + \xi^2$. Here $2 + 2\alpha + \alpha^2 = 0$. Thus,

$$\alpha(2 + 2\alpha + \alpha^2) = 0 \Rightarrow \alpha + \alpha^3 = -\alpha - 2\alpha^2 \equiv 2\alpha + \alpha^2 = -2 \bmod 3$$

$$\alpha + \alpha^3 \equiv -2 \Rightarrow \alpha(\alpha + \alpha^3) \equiv -2\alpha \Rightarrow \alpha^4 \equiv -2\alpha - \alpha^2 = 2 \bmod 3$$

Consequently, $\xi^2 - (\alpha + \alpha^3)\xi + \alpha^4 = 2 + 2\xi + \xi^2$ and $P_2(\xi) = (\xi - \alpha)(\xi - \alpha^3)$ is satisfied.

As a conclusion, the relation

$$P_2(\xi) = (\xi - \alpha)(\xi - \alpha^3)$$

works for each of the choices made for $P_2(\xi)$ in $\mathbb{F}_3[\xi]$.

2.5.5.4. *Example:* $\mathbb{GF}(3^3)$

For $\mathbb{GF}(3^3)$, i.e. $p = 3$ and $m = 3$, we can take

$$P_3(\xi) = 1 + 2\xi + \xi^3$$

as a primitive (and thus prime) polynomial in $\mathbb{F}_3[\xi]$. Let α be a root of $P_3(\xi)$. Thus, $1 + 2\alpha + \alpha^3 = 0$, and we may question whether $\alpha^p = \alpha^3$ and $\alpha^{p^2} = \alpha^9$ are the other roots of $P_3(\xi)$. Indeed, we have

$$1 + 2(\alpha^3) + (\alpha^3)^3 = 1 + 2\alpha^3 + \alpha^9$$

and

$$\alpha^9 = (\alpha^3)^3 = (-1 - 2\alpha)^3 \equiv (2 + \alpha)^3 \equiv 8 + \alpha^3 \equiv 2 + \alpha^3 \bmod 3$$

so that

$$1 + 2\alpha^3 + \alpha^9 = 1 + 2\alpha^3 + 2 + \alpha^3 \equiv 0 \bmod 3$$

Consequently, α^3 is a root of $P_3(\xi)$. Similarly, is α^9 a root of $P_3(\xi)$? We have

$$1 + 2(\alpha^9) + (\alpha^9)^3 = 1 + 2\alpha^9 + \alpha^{27}$$

From $\alpha^{p^m} \equiv \alpha \bmod p$, we get $\alpha^{27} \equiv \alpha \bmod 3$. Thus,

$$\begin{aligned} 1 + 2\alpha^9 + \alpha^{27} &\equiv 1 + 2(2 + \alpha^3) + \alpha = 5 + \alpha + 2\alpha^3 \\ &\equiv -1 - 2\alpha - \alpha^3 = 0 \bmod 3 \end{aligned}$$

Consequently, α^9 is a root of $P_3(\xi)$.

As another example, we can take

$$P_3(\xi) = 2 + 2\xi + \xi^3$$

as a prime (but not primitive) polynomial. It can be proved that this polynomial admits the roots α^3 and α^9 if α is a root of $P_3(\xi)$.

As an alternative proof of the preceding results, we can check that

$$P(\xi) = (\xi - \alpha)(\xi - \alpha^3)(\xi - \alpha^9)$$

is congruent to $P_3(\xi)$ modulo 3 with $P_3(\xi) = 1 + 2\xi + \xi^3$ or $2 + 2\xi + \xi^3$ (α is a root of $1 + 2\xi + \xi^3$ or $2 + 2\xi + \xi^3$). A simple development of $P(\xi)$ yields

$$P(\xi) = \xi^3 - (\alpha + \alpha^3 + \alpha^9)\xi^2 + (\alpha^4 + \alpha^{10} + \alpha^{12})\xi - \alpha^{13}$$

It is a matter of long-winded calculation to verify that

$$\alpha + \alpha^3 + \alpha^9 \equiv \begin{cases} 0 \text{ if } 1 + 2\alpha + \alpha^3 = 0 \\ \\ 0 \text{ if } 2 + 2\alpha + \alpha^3 = 0 \end{cases}$$

$$\alpha^4 + \alpha^{10} + \alpha^{12} \equiv \begin{cases} 2 \text{ if } 1 + 2\alpha + \alpha^3 = 0 \\ \\ 2 \text{ if } 2 + 2\alpha + \alpha^3 = 0 \end{cases}$$

$$\alpha^{13} \equiv \begin{cases} -1 \text{ if } 1 + 2\alpha + \alpha^3 = 0 \\ \\ -2 \text{ if } 2 + 2\alpha + \alpha^3 = 0 \end{cases}$$

modulo 3. We thus obtain

$$P(\xi) \equiv \begin{cases} 1 + 2\xi + \xi^3 \bmod 3 \text{ if } 1 + 2\alpha + \alpha^3 = 0 \\ 2 + 2\xi + \xi^3 \bmod 3 \text{ if } 2 + 2\alpha + \alpha^3 = 0 \end{cases}$$

so that $P(\xi) \equiv P_3(\xi)$ modulo 3 in both cases.

2.5.5.5. *Remarks*

At this stage, it is important to recall how the operations (addition and multiplication) are effected between the p^m elements of $\mathbb{GF}(p^m)$ with p prime and $m \geq 2$:

– when the elements of $\mathbb{GF}(p^m) = \mathbb{F}_p[\xi]/\langle P_m(\xi)\rangle$ are represented by classes of polynomials $x_0 + x_1\alpha + \cdots + x_{m-1}\alpha^{m-1}$ (α being a root of $P_m(\xi)$), the addition table for $(\mathbb{GF}(p^m), +)$ is easy to produce via sums of polynomials modulo p, and the multiplication table for $(\mathbb{GF}(p^m)^*, \times)$ can be obtained via products of polynomials modulo p and modulo the irreducible polynomial $P_m(\xi)$ (i.e. by taking into account $P_m(\alpha) = 0$);

– when the elements of $\mathbb{GF}(p^m) = \mathbb{F}_p[\xi]/\langle P_m(\xi)\rangle$ are taken in the form $0, \alpha, \cdots, \alpha^{p^m-1}$ (α being a primitive element of $\mathbb{GF}(p^m)$), the multiplication table for $(\mathbb{GF}(p^m)^*, \times)$ follows via products $\alpha^k\alpha^\ell = \alpha^{k+\ell}$ of monomials where the powers are added modulo $p^m - 1$, and the addition table for $(\mathbb{GF}(p^m), +)$ can be derived via sums $\alpha^k + 0$ and sums $\alpha^k + \alpha^\ell$ of monomials modulo p and modulo the use of $P_m(\alpha) = 0$ ($k, \ell = 1, 2, \cdots, p^m - 1$).

2.6. The application trace for a Galois field

2.6.1. *Trace of an element*

2.6.1.1. *Trace as a finite sum*

DEFINITION 2.7.– The *trace* of an element x of the field $\mathbb{GF}(p^m)$ with p prime and m positive integer, noted $\mathrm{Tr}(x)$, is defined by

$$\mathrm{Tr}(x) = \sum_{j=0}^{m-1} x^{p^j} = x^{p^0} + x^{p^1} + \cdots + x^{p^{m-1}} = x + x^p + \cdots + x^{p^{m-1}}$$

As a particular case, for $m = 1$, i.e. for $\mathbb{GF}(p) = \mathbb{F}_p = \mathbb{Z}_p$, $\mathrm{Tr}(x)$ is nothing but x ($\mathrm{Tr}(x) = x$).

In the special case where the element x of $\mathbb{GF}(p^m) = \mathbb{F}_p[\xi]/\langle P_m(\xi)\rangle$ is a root α of the prime polynomial

$$P_m(\xi) = c_0 + c_1\xi + \cdots + c_{m-1}\xi^{m-1} + \xi^m$$

we have

$$\mathrm{Tr}(\alpha) = -c_{m-1}$$

which is an element of $\mathbb{F}_p$.

2.6.1.2. *Trace as a map of $\mathbb{GF}(p^m)$ onto $\mathbb{F}_p$*

PROPOSITION 2.25.– The correspondence $x \mapsto \mathrm{Tr}(x)$ defines an application

$$\begin{aligned}\mathrm{Tr} : \mathbb{GF}(p^m) &\to \mathbb{F}_p \\ x &\mapsto \mathrm{Tr}(x)\end{aligned}$$

Such an application is surjective.

ELEMENT OF PROOF.– The fact that $\mathrm{Tr}(x)$ belongs to the prime field $\mathbb{F}_p$ follows from $[\mathrm{Tr}(x)]^p = \mathrm{Tr}(x)$ (remember that an element a in $\mathbb{GF}(p^m)$ belongs to $\mathbb{F}_p$ if and only if $a^p = a$). □

2.6.2. *Frobenius automorphism*

2.6.2.1. *The Frobenius map*

PROPOSITION 2.26.– The map

$$\sigma : \mathbb{GF}(p^m) \to \mathbb{GF}(p^m)$$
$$x \mapsto \sigma(x) = x^p$$

defines an automorphism of $\mathbb{GF}(p^m)$ called *Frobenius automorphism* of $\mathbb{GF}(p^m)$ over $\mathbb{F}_p$.

PROOF.– It is enough to prove that

$$\sigma(x+y) = \sigma(x) + \sigma(y), \quad \sigma(x \times y) = \sigma(x) \times \sigma(y)$$

for all x and y in $\mathbb{GF}(p^m)$. □

By defining $\sigma^1, \sigma^2, \cdots, \sigma^m$ via

$$\sigma^1(x) = \sigma(x), \quad \sigma^2(x) = \sigma(\sigma^1(x)), \quad \cdots, \quad \sigma^m(x) = \sigma(\sigma^{m-1}(x))$$

we have

$$\sigma^1(x) = x^{p^1}, \quad \sigma^2(x) = x^{p^2}, \quad \cdots, \quad \sigma^m(x) = x^{p^m}$$

with $\sigma^m(x) = x = \sigma^0(x)$. The powers $x^{p^1}, x^{p^2}, \cdots, x^{p^m} = x$ are called the *Galois conjugates* of $x \in \mathbb{GF}(p^m)$ with respect to $\mathbb{F}_p$. The Frobenius automorphism maps the Galois conjugates of $x \in \mathbb{GF}(p^m)$ to each other.

2.6.2.2. *Remarks*

– If α is a primitive element of $\mathbb{GF}(p^m)$, then all its Galois conjugates, with respect to $\mathbb{F}_p$, are primitive elements of $\mathbb{GF}(p^m)$ too.

– We have

$$\forall a \in \mathbb{F}_p \subset \mathbb{GF}(p^m) : \sigma(a) = a^p \equiv a \bmod p$$

Therefore, the elements of the prime field $\mathbb{F}_p$ are Galois self-conjugates (σ leaves all the elements of $\mathbb{F}_p$ fixed).

– Note that $\mathrm{Tr}(x)$ can be written as

$$\mathrm{Tr}(x) = \sum_{k=1}^{m} \sigma^k(x) = \sigma^1(x) + \sigma^2(x) + \cdots + \sigma^{m-1}(x) + \sigma^m(x)$$

or equivalently,

$$\mathrm{Tr}(x) = \sum_{j=0}^{m-1} \sigma^j(x) = \sigma^0(x) + \sigma^1(x) + \cdots + \sigma^{m-2}(x) + \sigma^{m-1}(x)$$

in terms of the Frobenius map so that $\mathrm{Tr}(x)$ is the sum of all the conjugates of x.

– The m maps

$$\sigma^1, \quad \sigma^2, \quad \cdots, \quad \sigma^m$$

are distinct automorphisms of $\mathbb{GF}(p^m)$ over $\mathbb{F}_p$. It is clear that the automorphisms σ^i $(i = 1, 2, \cdots, m)$ are the elements of a cyclic group with respect to the composition of maps (σ^m or σ^0 is the identity of the group). This cyclic group, of order m, generated by the Frobenius map σ is called the *Galois group* of $\mathbb{GF}(p^m)$ over $\mathbb{F}_p$.

2.6.2.3. *Example:* $\mathbb{GF}(2^2)$

We have

$$\sigma(x) = x^2, \quad \sigma(x^2) = x$$

for any element x of $\mathbb{GF}(2^2)$. The Galois group of $\mathbb{GF}(2^2)$ over $\mathbb{F}_2$ is isomorphic to the cyclic group C_2.

2.6.2.4. *Example:* $\mathbb{GF}(2^3)$

We have

$$\sigma(x) = x^2, \quad \sigma(x^2) = x^4, \quad \sigma(x^3) = x^6$$
$$\sigma(x^4) = x, \quad \sigma(x^5) = x^3, \quad \sigma(x^6) = x^5$$

for any element x of $\mathbb{GF}(2^3)$. The Galois group of $\mathbb{GF}(2^3)$ over $\mathbb{F}_2$ is isomorphic to the cyclic group C_3.

2.6.3. *Elementary properties of the trace*

2.6.3.1. *Four properties*

PROPOSITION 2.27.– The following properties

– Property 1:

$$\forall x \in \mathbb{GF}(p^m) : \mathrm{Tr}(-x) = -\mathrm{Tr}(x)$$

– Property 2:

$$\forall x \in \mathbb{GF}(p^m), \forall y \in \mathbb{GF}(p^m) : \mathrm{Tr}(xy) = \mathrm{Tr}(yx)$$

– Property 3:

$$\forall a \in \mathbb{F}_p \subset \mathbb{GF}(p^m) : \mathrm{Tr}(a) \equiv ma \bmod p$$

– Property 4:

$$\forall x \in \mathbb{GF}(p^m) : \mathrm{Tr}(x^p) \equiv \mathrm{Tr}(x) \equiv [\mathrm{Tr}(x)]^p \bmod p$$

and more generally,

$$\forall x \in \mathbb{GF}(p^m) : \mathrm{Tr}(x^{p^n}) \equiv \mathrm{Tr}(x) \equiv [\mathrm{Tr}(x)]^p \bmod p, \; n = 1, 2, \cdots, m$$

hold for any field $\mathbb{GF}(p^m)$ with p prime and $m \geq 1$.

PROOF.– Properties 1 and 2 are evident. For Property 3, let a be an element of $\mathbb{F}_p \subset \mathbb{GF}(p^m)$. Then

$$\begin{aligned} \mathrm{Tr}(a) &= a + a^p + \cdots + a^{p^{m-1}} \\ &\equiv a + a + \cdots + a \Rightarrow \mathrm{Tr}(a) \equiv ma \bmod p \end{aligned}$$

(remember that $a^p \equiv a \bmod p$ for $a \in \mathbb{F}_p \subset \mathbb{GF}(p^m)$) so that if m is a multiple of p, then $\mathrm{Tr}(a) \equiv 0 \bmod p$. For Property 4, if $x \in \mathbb{GF}(p^m)$ then

$$\begin{aligned} \mathrm{Tr}(x^p) &= (x^p)^{p^0} + (x^p)^{p^1} + \cdots + (x^p)^{p^{m-2}} + (x^p)^{p^{m-1}} \\ &\equiv x^{p^1} + x^{p^2} + \cdots + x^{p^{m-1}} + x \\ &= \mathrm{Tr}(x) \bmod p \end{aligned}$$

(remember that $x^{p^m} \equiv x$ for $x \in \mathbb{GF}(p^m)$). Similarly, it can be shown that

$$\mathrm{Tr}(x^{p^n}) \equiv \mathrm{Tr}(x) \bmod p, \quad n = 2, 3, \cdots, m-1$$

Furthermore,

$$\begin{aligned}
[\mathrm{Tr}(x)]^p &= (x + x^p + \cdots + x^{p^{m-2}} + x^{p^{m-1}})^p \\
&\equiv x^p + (x^p)^p + \cdots + (x^{p^{m-2}})^p + (x^{p^{m-1}})^p \\
&\equiv x^p + x^{p^2} + \cdots + x^{p^{m-1}} + x \\
&= \mathrm{Tr}(x) \bmod p
\end{aligned}$$

The congruence $\mathrm{Tr}(x^{p^n}) \equiv \mathrm{Tr}(x) \bmod p$ with $n = 1, 2, \cdots, m-1$ shows that all Galois conjugates have the same trace. The congruence $[\mathrm{Tr}(x)]^p \equiv \mathrm{Tr}(x) \bmod p$ shows that $\mathrm{Tr}(x)$ belongs to the prime field $\mathbb{F}_p$. □

2.6.3.2. *Notations*

From now on and when there is no ambiguity from the context, for the sake of simplicity, we will often use the sign = in place of the sign ≡. This means that some equalities should be understood modulo some prime number and/or by taking into account some prime polynomial.

2.6.3.3. *Example:* $\mathbb{GF}(2^2) = \mathbb{F}_2[\xi]/\langle 1 + \xi + \xi^2 \rangle$

For the field $\mathbb{GF}(2^2)$, we have

$$\forall x \in \mathbb{GF}(2^2) : \mathrm{Tr}(x) \equiv \mathrm{Tr}(x^2) \equiv [\mathrm{Tr}(x)]^2 = x + x^2 \bmod 2$$

In particular,

$$\mathrm{Tr}(0) = 0, \ \mathrm{Tr}(1) \equiv 0, \ \mathrm{Tr}(\alpha) = \alpha + \alpha^2 \equiv 1, \quad \mathrm{Tr}(\alpha^2) = \alpha^2 + \alpha^4 \equiv 1$$

modulo 2, where the primitive element α is such that $1 + \alpha + \alpha^2 = 0$.

Another expression of $\mathrm{Tr}(x)$ is obtained by using

$$x = x_0 + x_1\alpha, \quad x_0, x_1 \in \mathbb{F}_2$$

From $\mathrm{Tr}(x) = x + x^2$, we obtain

$$\begin{aligned}\mathrm{Tr}(x) &= x_0 + x_1\alpha + (x_0 + x_1\alpha)^2 \equiv x_0 + x_1\alpha + x_0^2 + x_1^2\alpha^2 \\ &\equiv x_1(\alpha + \alpha^2) \equiv x_1\end{aligned}$$

modulo 2 ($x_0^2 = x_0$ and $x_1^2 = x_1$). The result $\mathrm{Tr}(x) = x_1$ emphasizes the fact that the trace operation is a $\mathbb{GF}(2^2) \to \mathbb{F}_2$ surjection.

2.6.3.4. *Example:* $\mathbb{GF}(2^3) = \mathbb{F}_2[\xi]/\langle 1 + \xi + \xi^3 \rangle$

For the field $\mathbb{GF}(2^3)$, we obtain

$$\forall x \in \mathbb{GF}(2^3) : \mathrm{Tr}(x) = \mathrm{Tr}(x^2) = [\mathrm{Tr}(x)]^2 = x + x^2 + x^4$$

and the possible values of $\mathrm{Tr}(x)$ *a priori* are 0 and 1. In fact, a direct calculation of $x + x^2 + x^4$ for the various elements x of $\mathbb{GF}(2^3) = \mathbb{F}_2[\xi]/\langle 1 + \xi + \xi^3 \rangle$ yields

$$\begin{aligned}&\mathrm{Tr}(0) = \mathrm{Tr}(\alpha) = \mathrm{Tr}(\alpha^2) = \mathrm{Tr}(\alpha^4) = 0 \\ &\mathrm{Tr}(\alpha^3) = \mathrm{Tr}(\alpha^5) = \mathrm{Tr}(\alpha^6) = \mathrm{Tr}(\alpha^7) = 1\end{aligned}$$

in accordance with $\mathrm{Tr}(x) \in \mathbb{F}_2$ for any $x \in \mathbb{GF}(2^3)$ (α is a primitive element, root of $1 + \xi + \xi^3$).

By writing x as

$$x = x_0 + x_1\alpha + x_2\alpha^2, \quad x_0, x_1, x_2 \in \mathbb{F}_2$$

with α such that $1 + \alpha + \alpha^3 = 0$, formula $\mathrm{Tr}(x) = x + x^2 + x^4$ leads to

$$\begin{aligned}\mathrm{Tr}(x) &= x_0 + x_1\alpha + x_2\alpha^2 + (x_0 + x_1\alpha + x_2\alpha^2)^2 \\ &\quad + (x_0 + x_1\alpha + x_2\alpha^2)^4 \\ &= x_0 + x_1\alpha + x_2\alpha^2 + x_0^2 + x_1^2\alpha^2 + x_2^2\alpha^4 + x_0^4 + x_1^4\alpha^4 + x_2^4\alpha^8 \\ &= x_0 + x_1(\alpha + \alpha^2 + \alpha^4) + x_2(\alpha^2 + \alpha^4 + \alpha^8) \\ &= x_0 + x_1\alpha(1 + \alpha + \alpha^3) + x_2(\alpha^2 + \alpha^4 + \alpha) \\ &= x_0 + (x_1 + x_2)\alpha(1 + \alpha + \alpha^3)\end{aligned}$$

modulo 2. Finally,

$$\mathrm{Tr}(x) = x_0$$

in agreement with the fact that the trace operation is here a $\mathbb{GF}(2^3) \to \mathbb{F}_2$ surjection. By using the expressions

$$0,\ \alpha,\ \alpha^2,\ 1+\alpha = \alpha^3,\ \alpha+\alpha^2 = \alpha^4,\ 1+\alpha+\alpha^2 = \alpha^5$$

$$1+\alpha^2 = \alpha^6,\ 1 = \alpha^7$$

for $x = x_0+x_1\alpha+x_2\alpha^2$ written as powers of the primitive element α, formula $\mathrm{Tr}(x) = x_0$ gives the values obtained above for $\mathrm{Tr}(0)$, $\mathrm{Tr}(\alpha)$, $\cdots$, $\mathrm{Tr}(\alpha^7)$.

2.6.3.5. *Example:* $\mathbb{GF}(2^3) = \mathbb{F}_2[\xi]/\langle 1+\xi^2+\xi^3 \rangle$

Here we take

$$P_3(\xi) = 1+\xi^2+\xi^3$$

as primitive polynomial. Let α be a root of $P_3(\xi)$. The elements

$$x = x_0 + x_1\alpha + x_2\alpha^2, \quad x_0, x_1, x_2 \in \mathbb{F}_2$$

of $\mathbb{GF}(2^3)$ are then

$$0,\ \alpha,\ \alpha^2,\ \alpha^3 = 1+\alpha^2,\ \alpha^4 = 1+\alpha+\alpha^2$$

$$\alpha^5 = 1+\alpha,\ \alpha^6 = \alpha+\alpha^2,\ \alpha^7 = 1$$

On the one hand, the traces of these elements, calculated via the help of $\mathrm{Tr}(x) = x + x^2 + x^4$, read

$$\mathrm{Tr}(0) = \mathrm{Tr}(\alpha^3) = \mathrm{Tr}(\alpha^5) = \mathrm{Tr}(\alpha^6) = 0$$

$$\mathrm{Tr}(\alpha) = \mathrm{Tr}(\alpha^2) = \mathrm{Tr}(\alpha^4) = \mathrm{Tr}(\alpha^7) = 1$$

On the other hand, we obtain

$$\begin{aligned}\mathrm{Tr}(x) &= x_0 + x_1\alpha + x_2\alpha^2 + (x_0 + x_1\alpha + x_2\alpha^2)^2 \\ &\quad + (x_0 + x_1\alpha + x_2\alpha^2)^4 \\ &= x_0 + x_1\alpha + x_2\alpha^2 + x_0^2 + x_1^2\alpha^2 + x_2^2\alpha^4 + x_0^4 + x_1^4\alpha^4 + x_2^4\alpha^8 \\ &= x_0 + x_1(\alpha+\alpha^2+\alpha^4) + x_2(\alpha^2+\alpha^4+\alpha^8) \\ &= x_0 + x_1\alpha(1+\alpha+\alpha^3) + x_2(\alpha^2+\alpha^4+\alpha) \\ &= x_0 + (x_1+x_2)\alpha(1+\alpha+\alpha^3)\end{aligned}$$

$$= x_0 + (x_1 + x_2)\alpha(1 + \alpha + 1 + \alpha^2)$$
$$= x_0 + (x_1 + x_2)\alpha(\alpha + \alpha^2)$$
$$= x_0 + (x_1 + x_2)(\alpha^2 + \alpha^3)$$

and finally,

$$\text{Tr}(x) = x_0 + x_1 + x_2$$

to be compared with the result, $\text{Tr}(x) = x_0$, obtained in 2.6.3.4 for the extension $\mathbb{GF}(2^3) = \mathbb{F}_2[\xi]/\langle 1 + \xi + \xi^3 \rangle$.

As a conclusion, Examples 2.6.3.4 and 2.6.3.5 show that the expression of the trace of a generic element of $\mathbb{GF}(p^m) = \mathbb{F}_p[\xi]/\langle P_m(\xi) \rangle$ depends on the field extension (i.e. on the prime polynomial $P_m(\xi)$) used for constructing $\mathbb{GF}(p^m)$.

2.6.3.6. *Example:* $\mathbb{GF}(2^4) = \mathbb{F}_2[\xi]/\langle 1 + \xi + \xi^4 \rangle$

From

$$\forall x \in \mathbb{GF}(2^4) : \text{Tr}(x) = x + x^2 + x^4 + x^8$$

we obtain

$$\text{Tr}(0) = \text{Tr}(1) = \text{Tr}(\alpha) = \text{Tr}(\alpha^2) = 0$$
$$\text{Tr}(\alpha^4) = \text{Tr}(\alpha^5) = \text{Tr}(\alpha^8) = \text{Tr}(\alpha^{10}) = 0$$
$$\text{Tr}(\alpha^3) = \text{Tr}(\alpha^6) = \text{Tr}(\alpha^7) = \text{Tr}(\alpha^9) = 1$$
$$\text{Tr}(\alpha^{11}) = \text{Tr}(\alpha^{12}) = \text{Tr}(\alpha^{13}) = \text{Tr}(\alpha^{14}) = 1$$

where the primitive element α is a solution of $1 + \xi + \xi^4 = 0$. Furthermore, it can be shown that

$$\text{Tr}(x) = x_3$$

for $x = x_0 + x_1\alpha + x_2\alpha^2 + x_3\alpha^3$ in $\mathbb{F}_2[\xi]/\langle 1 + \xi + \xi^4 \rangle$.

2.6.3.7. *Example:* $\mathbb{GF}(3^2)$

Here $p = 3$ and $m = 2$. The generic element x of $\mathbb{GF}(3^2)$ reads

$$x = x_0 + x_1\alpha, \quad x_0 \in \mathbb{F}_3, \quad x_1 \in \mathbb{F}_3$$

where α is a root of some prime polynomial $P_2(\xi)$ in $\mathbb{F}_3[\xi]$. Then, we have

$$\mathrm{Tr}(x) = x + x^3$$

that leads to

$$\begin{aligned}\mathrm{Tr}(x) &= x_0 + x_1\alpha + (x_0 + x_1\alpha)^3 \\ &= x_0 + x_1\alpha + x_0^3 + x_1^3\alpha^3 \\ &= 2x_0 + x_1(\alpha + \alpha^3)\end{aligned}$$

since $x_0^3 = x_0$ and $x_1^3 = x_1$ in $\mathbb{F}_3$. There are three possibilities for $P_2(\xi)$ in $\mathbb{F}_3[\xi]$: $1 + \xi^2$ or $2 + \xi + \xi^2$ or $2 + 2\xi + \xi^2$. This yields

$$\mathrm{Tr}(x) = \begin{cases} 2x_0 \text{ for } P_2(\xi) = 1 + \xi^2 \\ 2x_0 + 2x_1 \text{ for } P_2(\xi) = 2 + \xi + \xi^2 \\ 2x_0 + x_1 \text{ for } P_2(\xi) = 2 + 2\xi + \xi^2 \end{cases}$$

Again, the latter result illustrates the fact that, for $x \in \mathbb{F}_p[\xi]/\langle P_m(\xi)\rangle$, the element $\mathrm{Tr}(x) \in \mathbb{F}_p$ depends on $P_m(\xi)$.

2.6.3.8. *Necessary and sufficient condition for* $\mathrm{Tr}(x) = 0$

PROPOSITION 2.28.– For x in $\mathbb{GF}(p^m)$ with $m \geq 1$, we have $\mathrm{Tr}(x) = 0$ if and only if there exists an element y in $\mathbb{GF}(p^m)$ such that $x = y - y^p$.

ELEMENT OF PROOF.– We limit ourselves to the sufficient condition of this theorem due to Hilbert. Assuming $x = y - y^p$, we get

$$\begin{aligned}\mathrm{Tr}(x) &= \mathrm{Tr}(y - y^p) \\ &= y - y^p + (y - y^p)^p + (y - y^p)^{p^2} + \cdots + (y - y^p)^{p^{m-1}} \\ &= y - y^p + y^p - y^{p^2} + y^{p^2} - y^{p^3} + \cdots + y^{p^{m-1}} - y^{p^m} \\ &= y - y^{p^m}\end{aligned}$$

Thus, $\mathrm{Tr}(x) = 0$. □

2.6.4. *Linearity of the trace*

2.6.4.1. *Linearity*

PROPOSITION 2.29.– For $\mathbb{GF}(p^m)$ with $m \geq 1$, two important properties of the trace are

$$\forall x \in \mathbb{GF}(p^m),\ \forall y \in \mathbb{GF}(p^m) : \mathrm{Tr}(x+y) = \mathrm{Tr}(x) + \mathrm{Tr}(y)$$

and

$$\forall a \in \mathbb{F}_p,\ \forall x \in \mathbb{GF}(p^m) : \mathrm{Tr}(ax) = a\mathrm{Tr}(x)$$
$$\forall a \in \mathbb{F}_p \subset \mathbb{GF}(p^m) : \mathrm{Tr}(a) = ma$$

so that the trace operation is a linear operation on $\mathbb{F}_p$.

PROOF.– The proof easily follows from $(x+y)^p = x^p + y^p$ in $\mathbb{GF}(p^m)$ and $a^p = a$ in $\mathbb{F}_p$. The proof is given in detail below for the fields $\mathbb{GF}(2^2)$, $\mathbb{GF}(2^3)$ and $\mathbb{GF}(2^4)$. □

2.6.4.2. *Example: linearity for* $\mathbb{GF}(2^2)$

We have

$$\mathrm{Tr}(x+y) = x + y + (x+y)^2 = x + y + x^2 + y^2 = \mathrm{Tr}(x) + \mathrm{Tr}(y)$$

for $x \in \mathbb{GF}(2^2)$ and $y \in \mathbb{GF}(2^2)$. Furthermore,

$$\mathrm{Tr}(ax) = ax + (ax)^2 = ax + a^2x^2 = a(x + x^2) = a\mathrm{Tr}(x)$$

for $a \in \mathbb{F}_2$ and $x \in \mathbb{GF}(2^2)$. Note that

$$\forall x \in \mathbb{GF}(2^2) : x = x_0 + x_1\alpha \Rightarrow \mathrm{Tr}(x) = x_0\mathrm{Tr}(1) + x_1\mathrm{Tr}(\alpha)$$

(with $x_0, x_1 \in \mathbb{F}_2$) gives back

$$\mathrm{Tr}(x) = x_1$$

for $\mathbb{GF}(2^2) = \mathbb{F}_2[\xi]/\langle 1 + \xi + \xi^2 \rangle$ since $\mathrm{Tr}(1) = 0$ and $\mathrm{Tr}(\alpha) = 1$, where α is a root of $1 + \xi + \xi^2 = 0$ (see 2.6.3.3).

2.6.4.3. *Example: linearity for* $\mathbb{GF}(2^3)$

Similarly, we obtain

$$\begin{aligned}\mathrm{Tr}(x+y) &= x+y+(x+y)^2+(x+y)^4\\ &= x+y+x^2+y^2+x^4+y^4\\ &= \mathrm{Tr}(x)+\mathrm{Tr}(y)\end{aligned}$$

for $x \in \mathbb{GF}(2^3)$ and $y \in \mathbb{GF}(2^3)$, and

$$\begin{aligned}\mathrm{Tr}(ax) &= ax+(ax)^2+(ax)^4\\ &= ax+a^2x^2+a^4x^4\\ &= a(x+x^2+x^4)\\ &= a\mathrm{Tr}(x)\end{aligned}$$

for $a \in \mathbb{F}_2$ and $x \in \mathbb{GF}(2^3)$. Note that

$$\forall x \in \mathbb{GF}(2^3) : x = x_0 + x_1\alpha + x_2\alpha^2$$
$$\Rightarrow \ \mathrm{Tr}(x) = x_0\mathrm{Tr}(1) + x_1\mathrm{Tr}(\alpha) + x_2\mathrm{Tr}(\alpha^2)$$

(with $x_0, x_1, x_2 \in \mathbb{F}_2$) leads to

$$\mathrm{Tr}(x) = x_0$$

for $\mathbb{GF}(2^3) = \mathbb{F}_2[\xi]/\langle 1+\xi+\xi^3\rangle$ since $\mathrm{Tr}(1) = 1$, $\mathrm{Tr}(\alpha) = 0$ and $\mathrm{Tr}(\alpha^2) = 0$, where α is a root of $1+\xi+\xi^3 = 0$ (see 2.6.3.4).

2.6.4.4. *Example: linearity for* $\mathbb{GF}(2^4)$

Here we have $p = 2$ and $m = 4$. Thus,

$$\forall x \in \mathbb{GF}(2^4) : \mathrm{Tr}(x) = x + x^2 + x^4 + x^8$$

Calculations analogous to those for $\mathbb{GF}(2^2)$ and $\mathbb{GF}(2^3)$ show that the trace is a linear operation on $\mathbb{F}_2$.

The general element x of $\mathbb{GF}(2^4)$ can be written as

$$x = x_0 + x_1\alpha + x_2\alpha^2 + x_3\alpha^3, \quad x_0, x_1, x_2, x_3 \in \mathbb{F}_2$$

where α stands for a root of a prime polynomial $P_4(\xi)$ of degree 4 in $\mathbb{F}_2[\xi]$. Then

$$\mathrm{Tr}(x) = x_0\mathrm{Tr}(1) + x_1\mathrm{Tr}(\alpha) + x_2\mathrm{Tr}(\alpha^2) + x_3\mathrm{Tr}(\alpha^3)$$

depends on α, i.e. on the chosen prime polynomial. For example, we can choose

$$P_4(\xi) = 1 + \xi + \xi^4 \text{ or } 1 + \xi^3 + \xi^4 \text{ or } 1 + \xi + \xi^2 + \xi^3 + \xi^4$$

This leads to the following results.

– For $\mathbb{GF}(2^4) = \mathbb{F}_2[\xi]/\langle 1 + \xi + \xi^4\rangle$, we get

$$\mathrm{Tr}(1) = \mathrm{Tr}(\alpha) = \mathrm{Tr}(\alpha^2) = 0, \quad \mathrm{Tr}(\alpha^3) = 1$$

with α solution of $1 + \xi + \xi^4 = 0$ so that

$$\forall x \in \mathbb{F}_2[\xi]/\langle 1 + \xi + \xi^4\rangle : \mathrm{Tr}(x) = x_3$$

– For $\mathbb{GF}(2^4) = \mathbb{F}_2[\xi]/\langle 1 + \xi^3 + \xi^4\rangle$, we obtain

$$\mathrm{Tr}(1) = 0, \quad \mathrm{Tr}(\alpha) = \mathrm{Tr}(\alpha^2) = \mathrm{Tr}(\alpha^3) = 1$$

with α solution of $1 + \xi^3 + \xi^4 = 0$ so that

$$\forall x \in \mathbb{F}_2[\xi]/\langle 1 + \xi^3 + \xi^4\rangle : \mathrm{Tr}(x) = x_1 + x_2 + x_3$$

– For $\mathbb{GF}(2^4) = \mathbb{F}_2[\xi]/\langle 1 + \xi + \xi^2 + \xi^3 + \xi^4\rangle$, we have

$$\mathrm{Tr}(1) = 0, \quad \mathrm{Tr}(\alpha) = \mathrm{Tr}(\alpha^2) = \mathrm{Tr}(\alpha^3) = 1$$

with α solution of $1 + \xi + \xi^2 + \xi^3 + \xi^4 = 0$ so that

$$\forall x \in \mathbb{F}_2[\xi]/\langle 1 + \xi + \xi^2 + \xi^3 + \xi^4\rangle : \mathrm{Tr}(x) = x_1 + x_2 + x_3$$

The latter three results show again that the trace of an element of $\mathbb{F}_p[\xi]/\langle P_m(\xi)\rangle$ may depend on the prime polynomial $P_m(\xi)$.

2.6.5. *Trace in terms of the roots of a prime polynomial*

2.6.5.1. *A formula for the trace*

PROPOSITION 2.30.– Let us consider the field $\mathbb{GF}(p^m) = \mathbb{F}_p[\xi]/\langle P_m(\xi)\rangle$ and let $\alpha_0, \alpha_1, \cdots, \alpha_{m-1}$ be the m distinct roots of the monic irreducible polynomial $P_m(\xi)$ in $\mathbb{F}_p[\xi]$ with $m \geq 2$. Then

$$\forall x \in \mathbb{F}_p[\xi]/\langle P_m(\xi)\rangle : \mathrm{Tr}(x) = \sum_{k=0}^{m-1} x(\alpha_k)$$

where $x(\alpha_k)$ is the expression of x in terms of the root α_k ($k = 0, 1, \cdots, m-1$) of the prime polynomial $P_m(\xi)$.

2.6.5.2. *Example:* $\mathbb{GF}(2^2)$ *and* $\mathbb{GF}(2^3)$

– For $\mathbb{GF}(2^2) = \mathbb{F}_2[\xi]/\langle 1 + \xi + \xi^2\rangle$, we have

$$\alpha_0 + \alpha_1 = -1$$

Thus, the traces of the elements 0, 1, α, $1 + \alpha \equiv \alpha^2 \bmod 2$ of $\mathbb{GF}(2^2)$ are

$$\begin{aligned}
\mathrm{Tr}(0) &= 0(\alpha_0) + 0(\alpha_1) = 0 + 0 = 0\\
\mathrm{Tr}(1) &= 1(\alpha_0) + 1(\alpha_1) = 1 + 1 \equiv 0 \bmod 2\\
\mathrm{Tr}(\alpha) &= \alpha(\alpha_0) + \alpha(\alpha_1) = \alpha_0 + \alpha_1 = -1 \equiv 1 \bmod 2\\
\mathrm{Tr}(1+\alpha) &= (1+\alpha)(\alpha_0) + (1+\alpha)(\alpha_1) = 1 + \alpha_0 + 1 + \alpha_1 = 1
\end{aligned}$$

in agreement with section 2.6.3.3.

– For $\mathbb{GF}(2^3) = \mathbb{F}_2[\xi]/\langle 1 + \xi + \xi^3\rangle$, we have

$$\alpha_0 + \alpha_1 + \alpha_2 = 0$$

so that

$$\begin{aligned}
\mathrm{Tr}(0) &= 0 + 0 + 0 = 0\\
\mathrm{Tr}(1) &= 1 + 1 + 1 \equiv 1 \bmod 2\\
\mathrm{Tr}(\alpha^1) &= \alpha_0 + \alpha_1 + \alpha_2 = 0\\
\mathrm{Tr}(\alpha^2) &= \alpha_0^2 + \alpha_1^2 + \alpha_2^2 \equiv (\alpha_0 + \alpha_1 + \alpha_2)^2 = 0 \bmod 2
\end{aligned}$$

$$\mathrm{Tr}(\alpha^3) = \alpha_0^3 + \alpha_1^3 + \alpha_2^3 \equiv 1 + \alpha_0 + 1 + \alpha_1 + 1 + \alpha_2 \equiv 1 \bmod 2$$
$$\mathrm{Tr}(\alpha^4) = \alpha_0^4 + \alpha_1^4 + \alpha_2^4 \equiv (\alpha_0 + \alpha_1 + \alpha_2)^4 = 0 \bmod 2$$
$$\mathrm{Tr}(\alpha^5) = \alpha_0^5 + \alpha_1^5 + \alpha_2^5 \equiv \alpha_0^2 + \alpha_0^3 + \alpha_1^2 + \alpha_1^3 + \alpha_2^2 + \alpha_2^3 \equiv 1 \bmod 2$$
$$\mathrm{Tr}(\alpha^6) = \alpha_0^6 + \alpha_1^6 + \alpha_2^6 \equiv (\alpha_0^3 + \alpha_1^3 + \alpha_2^3)^2 \equiv 1 \bmod 2$$

in agreement with section 2.6.3.4.

2.7. Bases of a Galois field

2.7.1. *Generalities*

The field $\mathbb{GF}(p^m) = \mathbb{F}_p[\xi]/\langle P_m(\xi)\rangle$ contains p^m elements. Each element x of $\mathbb{GF}(p^m)$ can be written in the following way

$$x = x_0 + x_1\alpha + \cdots + x_{m-1}\alpha^{m-1}, \quad x_0, x_1, \cdots, x_{m-1} \in \mathbb{F}_p$$

where α is an element such that $P_m(\alpha) = 0$, $P_m(\xi)$ being a prime (i.e. monic + irreducible) polynomial in $\mathbb{F}_p[\xi]$. Therefore, the elements of $\mathbb{GF}(p^m)$ can be seen as the elements of a vector space, of dimension m, over the field $\mathbb{F}_p$. Then, the set $\{1, \alpha, \cdots, \alpha^{m-1}\}$ plays the role of a basis for the field $\mathbb{GF}(p^m)$ considered as a vector space over the field $\mathbb{F}_p$ (i.e. a $\mathbb{F}_p$-vector space). Note that the elements of $\mathbb{F}_p$ correspond to those of the field $\mathbb{GF}(p^m)$ with $x_1 = x_2 = \cdots = x_{m-1} = 0$.

One says that $\{\alpha^k \mid 0 \leq k \leq m-1\}$ is a basis for the field $\mathbb{GF}(p^m)$. Such a basis is called a *polynomial basis* in the sense that each element of $\mathbb{GF}(p^m)$ can be expressed (in a unique way) as a linear combination of the monomials $\alpha^0 = 1$, $\alpha^1 = \alpha$, $\cdots$, α^{m-1}.

In the limit case $m = 1$, the field $\mathbb{GF}(p) = \mathbb{F}_p$ can be considered as a one-dimensional vector space with basis $\{1\}$.

As an illustration, for the field $\mathbb{GF}(2^4)$, corresponding to $p = 2$ and $m = 4$, we have a vector space of dimension 4 over the field $\mathbb{F}_2$. The set $\{\alpha^0, \alpha^1, \alpha^2, \alpha^3\}$, where α is a root of a prime polynomial $P_4(\xi)$, constitutes a basis for $\mathbb{GF}(2^4)$.

More generally, a basis of the field $\mathbb{GF}(p^m)$ can be defined as follows.

2.7.2. *Field bases*

2.7.2.1. *Basis of a Galois field*

DEFINITION 2.8.– A basis of the Galois field $\mathbb{GF}(p^m)$ is a set $\{e_0, e_1, \cdots, e_{m-1}\}$ of m elements of $\mathbb{GF}(p^m)$ such that every element x of $\mathbb{GF}(p^m)$ can be developed as

$$x = \sum_{k=0}^{m-1} x_k e_k = x_0 e_0 + x_1 e_1 + \cdots + x_{m-1} e_{m-1}$$

where the x_k ($k = 0, 1, \cdots, m-1$) belong to $\mathbb{F}_p$.

2.7.2.2. *Example: bases of* $\mathbb{GF}(2^2) = \mathbb{F}_2[\xi]/\langle 1 + \xi + \xi^2 \rangle$

Let α be a root of $1 + \xi + \xi^2$. First, the basis

$$B_1 = \{1, \alpha\}$$

yields 0, 1, α and $1 + \alpha$ for the elements of the field $\mathbb{GF}(2^2)$. Second, another basis is

$$B_2 = \{\alpha, \alpha^2\}$$

for which the elements of $\mathbb{GF}(2^2)$ are 0, α, α^2 and $\alpha + \alpha^2$. Third, the set

$$B_3 = \{1, \alpha^2\}$$

constitutes a basis in which the elements of $\mathbb{GF}(2^2)$ are 0, 1, α^2 and $1 + \alpha^2$. The relations

$$\alpha^2 = 1 + \alpha, \quad \alpha + \alpha^2 = 1, \quad \alpha = 1 + \alpha^2$$

(which follow from $1 + \alpha + \alpha^2 = 0$ mod 2) make it possible to connect the three bases B_1, B_2 and B_3.

2.7.2.3. *Example: bases of* $\mathbb{GF}(2^3) = \mathbb{F}_2[\xi]/\langle 1+\xi^2+\xi^3\rangle$

Table 2.25 gives the elements of $\mathbb{GF}(2^3)$ in the bases

- $\{1, \alpha, \alpha^2\} = \{\alpha^7, \alpha^1, \alpha^2\}$
- $\{\alpha, \alpha^2, \alpha^4\} = \{\alpha^1, \alpha^2, \alpha^4\}$
- $\{\alpha, \alpha^2, 1+\alpha+\alpha^2\} = \{\alpha^1, \alpha^2, \alpha^7+\alpha^1+\alpha^2\}$

where α is a primitive element of $\mathbb{F}_2[\xi]/\langle 1+\xi^2+\xi^3\rangle$.

Elements of $\mathbb{GF}(2^3)$	Basis $\{\alpha^7, \alpha^1, \alpha^2\}$	Basis $\{\alpha^1, \alpha^2, \alpha^4\}$	Basis $\{\alpha^1, \alpha^2, \alpha^7+\alpha^1+\alpha^2\}$
0	[000]	[000]	[000]
α^1	[010]	[100]	[100]
α^2	[001]	[010]	[010]
α^3	[101]	[101]	[101]
α^4	[111]	[001]	[001]
α^5	[110]	[011]	[011]
α^6	[011]	[110]	[110]
$1=\alpha^7$	[100]	[111]	[111]

Table 2.25. *Elements of* $\mathbb{GF}(2^3) = \mathbb{F}_2[\xi]/\langle 1+\xi^2+\xi^3\rangle$ *in three different bases* $\{\alpha^a, \alpha^b, \alpha^c\}$ *(the element* $x\alpha^a + y\alpha^b + z\alpha^c$ *in the basis* $\{\alpha^a, \alpha^b, \alpha^c\}$ *is denoted as* $[xyz]$*; for instance, in the basis* $\{\alpha^7, \alpha^1, \alpha^2\}$*,* $[111]$ *stands for* $\alpha^4 \equiv 1+\alpha+\alpha^2$ *mod 2 and* $[011]$ *for* $\alpha^6 \equiv \alpha+\alpha^2$ *mod 2)*

2.7.2.4. *Necessary and sufficient condition for a basis*

PROPOSITION 2.31.– Let $\{x, y, \cdots, z\}$ be a set of m elements of a field $\mathbb{GF}(p^m) = \mathbb{F}_p[\xi]/\langle P_m(\xi)\rangle$. This set constitutes a basis of the vector space $\mathbb{GF}(p^m)$ over the field $\mathbb{F}_p$ if and only if the $m \times m$ matrix

$$\begin{pmatrix} x & y & \cdots & z \\ x^p & y^p & \cdots & z^p \\ \vdots & \vdots & \cdots & \vdots \\ x^{p^{m-1}} & y^{p^{m-1}} & \cdots & z^{p^{m-1}} \end{pmatrix}$$

has a determinant different from zero.

Note that the determinant is calculated modulo p and modulo the prime polynomial $P_m(\xi)$.

2.7.2.5. *Polynomial basis*

We have already encountered polynomial bases $\{1,\ \alpha,\ \cdots,\ \alpha^{m-1}\}$ of the vector space $\mathbb{GF}(p^m)$ over the field $\mathbb{F}_p$ where α is a root of some prime polynomial $P_m(\xi)$ in $\mathbb{F}_p[\xi]$. More precisely, we have the following definition.

DEFINITION 2.9.– The basis

$$B_{\text{pol}} = \{\alpha^0 = 1,\ \alpha^1 = \alpha,\ \cdots,\ \alpha^{m-1}\}$$

of $\mathbb{GF}(p^m) = \mathbb{F}_p[\xi]/\langle P_m(\xi)\rangle$ is called a polynomial basis (α is a root of the prime polynomial $P_m(\xi)$). In a polynomial basis, the elements of $\mathbb{GF}(p^m)$ are given by polynomials in the variable α of degree lower than or equal to $m-1$ with coefficients in $\mathbb{F}_p$.

Note that the prime polynomial $P_m(\xi)$ can be a primitive polynomial. In this case, α is a primitive element of $\mathbb{GF}(p^m)$.

2.7.2.6. *Example:* $\mathbb{GF}(2^2)$, $\mathbb{GF}(2^3)$, $\mathbb{GF}(3^2)$ *and* $\mathbb{GF}(3^3)$

As an illustration of Proposition 2.7.2.4, it can be checked that

– for $\mathbb{GF}(2^2) = \mathbb{F}_2[\xi]/\langle 1+\xi+\xi^2\rangle$, we have

$$\det\begin{pmatrix} 1 & \alpha \\ 1 & \alpha^2 \end{pmatrix} = 1 \Rightarrow \{1,\alpha\} \text{ is a polynomial basis}$$

– for $\mathbb{GF}(2^3) = \mathbb{F}_2[\xi]/\langle 1+\xi+\xi^3\rangle$, we have

$$\det\begin{pmatrix} 1 & \alpha & \alpha^2 \\ 1 & \alpha^2 & \alpha^4 \\ 1 & \alpha^4 & \alpha^8 \end{pmatrix} = 1 \Rightarrow \{1,\alpha,\alpha^2\} \text{ is a polynomial basis}$$

– for $\mathbb{GF}(3^2) = \mathbb{F}_3[\xi]/\langle 1+\xi^2\rangle$, we have

$$\det\begin{pmatrix} 1 & \alpha \\ 1 & \alpha^3 \end{pmatrix} = \alpha \Rightarrow \{1,\alpha\} \text{ is a polynomial basis}$$

– for $\mathbb{GF}(3^3) = \mathbb{F}_3[\xi]/\langle 2+2\xi+\xi^3\rangle$, we have

$$\det\begin{pmatrix} 1 & \alpha & \alpha^2 \\ 1 & \alpha^3 & \alpha^6 \\ 1 & \alpha^9 & \alpha^{18} \end{pmatrix} = 2 \Rightarrow \{1,\alpha,\alpha^2\} \text{ is a polynomial basis}$$

where in each case, α is a root of the relevant prime polynomial $P_m(\xi)$.

2.7.2.7. *Normal basis*

DEFINITION 2.10.– A basis

$$B_{\text{nor}} = \{e_0, e_1, \cdots, e_{m-1}\}$$

of the vector space $\mathbb{GF}(p^m)$ over the field $\mathbb{F}_p$ is said to be a *normal basis* if there exists an element x of the field $\mathbb{GF}(p^m)$ such that

$$e_0 = \sigma^0(x) = x^{p^0},\ e_1 = \sigma^1(x) = x^{p^1},\ \cdots,\ e_{m-1} = \sigma^{m-1}(x) = x^{p^{m-1}}$$

(up to a rearrangement of $e_0, e_1, \cdots, e_{m-1}$) where σ is the Frobenius automorphism of $\mathbb{GF}(p^m)$ over $\mathbb{F}_p$.

A normal basis consists of all the conjugates of a single element. Every field $\mathbb{GF}(p^m)$ possesses a normal basis (non-unique in general).

2.7.2.8. *Necessary and sufficient condition for a normal basis*

PROPOSITION 2.32.– Let x be an element of the field $\mathbb{GF}(p^m) = \mathbb{F}_p[\xi]/\langle P_m(\xi)\rangle$. The set

$$\{x, x^p, \cdots, x^{p^{m-1}}\}$$

is a normal basis of the vector space $\mathbb{GF}(p^m)$ over the field $\mathbb{F}_p$ if and only if the $m \times m$ matrix

$$\begin{pmatrix} x & x^p & \cdots & x^{p^{m-1}} \\ x^p & x^{p^2} & \cdots & x^{p^m} \\ \vdots & \vdots & \cdots & \vdots \\ x^{p^{m-1}} & x^{p^m} & \cdots & x^{p^{2m-2}} \end{pmatrix}$$

has a determinant different from zero.

PROOF.– This result is a simple corollary of Proposition 2.7.2.4. □

2.7.2.9. *Counter-example:* $\mathbb{GF}(2^3) = \mathbb{F}_2[\xi]/\langle 1 + \xi + \xi^3 \rangle$

The set

$$\{\alpha, \alpha^2, \alpha^4\}$$

where α is a primitive element of $\mathbb{GF}(2^3)$ does not constitute a normal basis of $\mathbb{F}_2[\xi]/\langle 1 + \xi + \xi^3 \rangle$. This follows from the fact that α, α^2 and α^4 are not linearly independent ($1 + \alpha + \alpha^3 = 0 \Rightarrow \alpha + \alpha^2 + \alpha^4 = 0$). Another way to obtain this result is to use Proposition 2.7.2.8. Then, we easily get

$$\det \begin{pmatrix} \alpha & \alpha^2 & \alpha^4 \\ \alpha^2 & \alpha^4 & \alpha^8 \\ \alpha^4 & \alpha^8 & \alpha^{16} \end{pmatrix} = 0$$

and we thus recover that

$$\{\alpha,\ \sigma(\alpha) = \alpha^2,\ \sigma^2(\alpha) = \alpha^4 = \alpha + \alpha^2\}$$

is not a normal basis of $\mathbb{F}_2[\xi]/\langle 1 + \xi + \xi^3 \rangle$.

2.7.2.10. *Example:* $\mathbb{GF}(2^3) = \mathbb{F}_2[\xi]/\langle 1 + \xi^2 + \xi^3 \rangle$

Let α be a root of $1 + \xi^2 + \xi^3$. Here, we get

$$\det \begin{pmatrix} \alpha & \alpha^2 & \alpha^4 \\ \alpha^2 & \alpha^4 & \alpha^8 \\ \alpha^4 & \alpha^8 & \alpha^{16} \end{pmatrix} = 1$$

Therefore,

$$\{\alpha,\ \sigma(\alpha) = \alpha^2,\ \sigma^2(\alpha) = \alpha^4 = 1 + \alpha + \alpha^2\}$$

is a normal basis of $\mathbb{F}_2[\xi]/\langle 1 + \xi^2 + \xi^3 \rangle$.

2.7.2.11. *Example: various bases of* $\mathbb{GF}(2^3) = \mathbb{F}_2[\xi]/\langle 1 + \xi + \xi^3 \rangle$

In this case, the polynomial basis is

$$B_{\text{pol}} = \{1, \alpha, \alpha^2\} = \{\alpha^7, \alpha^1, \alpha^2\}$$

where α is a root of $1 + \xi + \xi^3$. As alternative bases, we can take

$$B_{724} = \{\alpha^7, \alpha^2, \alpha^4\}, \quad B_{365} = \{\alpha^3, \alpha^6, \alpha^5\}$$

$$B_{435} = \{\alpha^4, \alpha^3, \alpha^5\}, \quad B_{645} = \{\alpha^6, \alpha^4, \alpha^5\}$$

Table 2.26 gives the elements of $\mathbb{GF}(2^3)$ in the bases $B_{\text{pol}} = B_{712}$, B_{724}, B_{365}, B_{435} and B_{645}.

Elements of $\mathbb{GF}(2^3)$	Basis $\{\alpha^7\alpha^1\alpha^2\}$	Basis $\{\alpha^7\alpha^2\alpha^4\}$	Basis $\{\alpha^3\alpha^6\alpha^5\}$	Basis $\{\alpha^4\alpha^3\alpha^5\}$	Basis $\{\alpha^6\alpha^4\alpha^5\}$
0	[000]	[000]	[000]	[000]	[000]
$1 = \alpha^7$	[100]	[100]	[111]	[101]	[011]
α^1	[010]	[011]	[011]	[111]	[101]
α^2	[001]	[010]	[101]	[011]	[111]
α^3	[110]	[111]	[100]	[010]	[110]
α^4	[011]	[001]	[110]	[100]	[010]
α^5	[111]	[101]	[001]	[001]	[001]
α^6	[101]	[110]	[010]	[110]	[100]

Table 2.26. *Elements of $\mathbb{GF}(2^3) = \mathbb{F}_2[\xi]/\langle 1+\xi+\xi^3\rangle$ in five different bases $\{\alpha^a, \alpha^b, \alpha^c\}$, abbreviated $\{\alpha^a\alpha^b\alpha^c\}$ (the element $x\alpha^a + y\alpha^b + z\alpha^c$ in the basis $\{\alpha^a\alpha^b\alpha^c\}$ is denoted as $[xyz]$; for instance, in the basis $\{\alpha^7\alpha^1\alpha^2\}$, $[011]$ stands for $\alpha^4 \equiv \alpha + \alpha^2$ mod 2 and $[111]$ for $\alpha^5 \equiv 1 + \alpha + \alpha^2$ mod 2)*

We have seen in 2.7.2.9 that the set

$$\{\alpha, \alpha^2, \alpha^4\}$$

does not constitute a normal basis of $\mathbb{F}_2[\xi]/\langle 1 + \xi + \xi^3\rangle$. Nevertheless, it is possible to find a normal basis for $\mathbb{F}_2[\xi]/\langle 1 + \xi + \xi^3\rangle$: it can be checked that the set

$$\{e_0 = 1 + \alpha,\ e_1 = (1 + \alpha)^2 = 1 + \alpha^2,\ e_2 = (1 + \alpha)^4 = 1 + \alpha^4\}$$

provides a normal basis. In this basis, the elements of $\mathbb{F}_2[\xi]/\langle 1 + \xi + \xi^3\rangle$ read

$$0,\ 1,\ \alpha^6 = e_1,\ \alpha^5 = e_2,\ \alpha^4 = e_0 + e_1,\ \alpha^3 = e_0$$

$$\alpha^2 = e_0 + e_2,\ \alpha = e_1 + e_2$$

with $0 = 0e_0 + 0e_1 + 0e_2$ and $1 = \alpha^7 = e_0 + e_1 + e_2$. Another normal basis for $\mathbb{F}_2[\xi]/\langle 1 + \xi + \xi^3 \rangle$ is

$$\{e_0 = 1 + \alpha + \alpha^2,\ e_1 = (1 + \alpha + \alpha^2)^2 = 1 + \alpha$$

$$e_2 = (1 + \alpha + \alpha^2)^4 = 1 + \alpha^2\}$$

In such a basis, the elements of $\mathbb{F}_2[\xi]/\langle 1 + \xi + \xi^3 \rangle$ are

$$0,\ 1,\ \alpha^6 = e_2,\ \alpha^5 = e_0,\ \alpha^4 = e_1 + e_2,\ \alpha^3 = e_1$$

$$\alpha^2 = e_0 + e_1,\ \alpha = e_0 + e_2$$

with $0 = 0e_0 + 0e_1 + 0e_2$ and $1 = \alpha^7 = e_0 + e_1 + e_2$. The latter two bases illustrate the fact that a Galois field may have more than one normal basis.

2.7.3. *Dual and self-dual bases*

2.7.3.1. *Dual bases*

PROPOSITION 2.33.– Given any basis $B = \{e_0, e_1, \cdots, e_{m-1}\}$ of the vector space $\mathbb{GF}(p^m)$ over the field $\mathbb{F}_p$, there is a unique basis $\tilde{B} = \{\tilde{e}_0, \tilde{e}_1, \cdots, \tilde{e}_{m-1}\}$ such that

$$\mathrm{Tr}(e_i \tilde{e}_j) = \delta(i, j), \quad i, j = 0, 1, \cdots, m-1$$

The basis $\tilde{B}$ is called the *dual basis* of B and vice versa.

2.7.3.2. *Self-dual basis*

DEFINITION 2.11.– If $\tilde{B} = B$, the basis B is referred to as a *self-dual basis*.

A given field $\mathbb{GF}(p^m)$ has a self-dual basis if either $p = 2$ or both p and m are odd. For p odd and m even, $\mathbb{GF}(p^m)$ does not have a self-dual basis.

2.7.3.3. *Example: dual bases of* $\mathbb{GF}(2^2)$, $\mathbb{GF}(2^3)$ *and* $\mathbb{GF}(2^4)$

– The polynomial basis $\{1, \alpha\}$ and the basis $\{\alpha^2, 1\}$ are dual bases of $\mathbb{GF}(2^2) = \mathbb{F}_2[\xi]/\langle 1 + \xi + \xi^2 \rangle$.

– The polynomial basis $\{1, \alpha, \alpha^2\}$ and the basis $\{1, \alpha^2, \alpha\}$ constitute a couple of dual bases of $\mathbb{GF}(2^3) = \mathbb{F}_2[\xi]/\langle 1 + \xi + \xi^3 \rangle$. The bases $\{\alpha^4, \alpha^3, \alpha^5\}$ and $\{\alpha^6, \alpha^4, \alpha^5\}$ form another couple of dual bases of the field $\mathbb{GF}(2^3) = \mathbb{F}_2[\xi]/\langle 1 + \xi + \xi^3 \rangle$.

– Table 2.27 gives couples of dual bases of the field $\mathbb{GF}(2^3) = \mathbb{F}_2[\xi]/\langle 1 + \xi^2 + \xi^3 \rangle$.

– The polynomial basis $\{1, \alpha, \alpha^2, \alpha^3\}$ and the basis $\{1 + \alpha^3, \alpha^2, \alpha, 1\}$ are dual bases of $\mathbb{GF}(2^4) = \mathbb{F}_2[\xi]/\langle 1 + \xi + \xi^4 \rangle$.

In each of the cases above, α is a root of the relevant primitive polynomial.

Basis of $\mathbb{F}_2[\xi]/\langle 1 + \xi^2 + \xi^3 \rangle$	Dual Basis of $\mathbb{F}_2[\xi]/\langle 1 + \xi^2 + \xi^3 \rangle$	Multiplication Factor
$\{\alpha^1, \alpha^5, \alpha^6\}$	$\{\alpha^7, \alpha^4, \alpha^5\}$	α^1
$\{\alpha^2, \alpha^3, \alpha^5\}$	$\{\alpha^7, \alpha^1, \alpha^3\}$	α^2
$\{\alpha^7, \alpha^2, \alpha^6\}$	$\{\alpha^4, \alpha^6, \alpha^3\}$	α^3
$\{\alpha^3, \alpha^4, \alpha^6\}$	$\{\alpha^6, \alpha^7, \alpha^2\}$	α^4
$\{\alpha^7, \alpha^1, \alpha^3\}$	$\{\alpha^2, \alpha^3, \alpha^5\}$	α^5
$\{\alpha^7, \alpha^4, \alpha^5\}$	$\{\alpha^1, \alpha^5, \alpha^6\}$	α^6
$\{\alpha^1, \alpha^2, \alpha^4\}$	$\{\alpha^1, \alpha^2, \alpha^4\}$	$\alpha^7 \equiv 1$

Table 2.27. *Couples of dual bases of the field $\mathbb{GF}(2^3) = \mathbb{F}_2[\xi]/\langle 1 + \xi^2 + \xi^3 \rangle$; each basis can be deduced from its dual basis via a global multiplication factor: as an illustration, the basis $\{\alpha^7, \alpha^1, \alpha^3\}$ follows from its dual basis $\{\alpha^2, \alpha^3, \alpha^5\}$ by multiplying each element of the dual basis by α^5 ($\alpha^2 \times \alpha^5 = \alpha^7 \equiv 1$, $\alpha^3 \times \alpha^5 = \alpha^8 \equiv \alpha^1$, $\alpha^5 \times \alpha^5 = \alpha^{10} \equiv \alpha^3$)*

2.7.3.4. *Example: self-dual bases of $\mathbb{GF}(2^2)$ and $\mathbb{GF}(2^3)$*

– The normal basis $\{\alpha, \alpha^2\}$ of $\mathbb{GF}(2^2) = \mathbb{F}_2[\xi]/\langle 1 + \xi + \xi^2 \rangle$ is self-dual since

$$\mathrm{Tr}(\alpha \times \alpha) = \mathrm{Tr}(\alpha^2 \times \alpha^2) = 1, \quad \mathrm{Tr}(\alpha \times \alpha^2) = \mathrm{Tr}(\alpha^2 \times \alpha) = 0$$

(For the sake of clarity, we sometimes re-introduce the $\times$ sign in order to emphasize the multiplication law.)

– For $\mathbb{GF}(2^3) = \mathbb{F}_2[\xi]/\langle 1 + \xi + \xi^3 \rangle$, we can show that

$$\mathrm{Tr}(\alpha^3 \times \alpha^3) = \mathrm{Tr}(\alpha^6 \times \alpha^6) = \mathrm{Tr}(\alpha^5 \times \alpha^5) = 1$$

$$\mathrm{Tr}(\alpha^3 \times \alpha^6) = \mathrm{Tr}(\alpha^6 \times \alpha^5) = \mathrm{Tr}(\alpha^5 \times \alpha^3) = 0$$

Therefore, the basis $\{\alpha^3, \alpha^6, \alpha^5\}$ of $\mathbb{GF}(2^3) = \mathbb{F}_2[\xi]/\langle 1 + \xi + \xi^3 \rangle$ is self-dual.

– Similarly, for $\mathbb{GF}(2^3) = \mathbb{F}_2[\xi]/\langle 1 + \xi^2 + \xi^3 \rangle$, we have

$$\text{Tr}(\alpha^1 \times \alpha^1) = \text{Tr}(\alpha^2 \times \alpha^2) = \text{Tr}(\alpha^4 \times \alpha^4) = 1$$

$$\text{Tr}(\alpha^1 \times \alpha^2) = \text{Tr}(\alpha^2 \times \alpha^4) = \text{Tr}(\alpha^4 \times \alpha^1) = 0$$

Therefore, the basis $\{\alpha^1, \alpha^2, \alpha^4\}$ of $\mathbb{GF}(2^3) = \mathbb{F}_2[\xi]/\langle 1 + \xi^2 + \xi^3 \rangle$ is self-dual. Another self-dual basis of the field $\mathbb{GF}(2^3) = \mathbb{F}_2[\xi]/\langle 1 + \xi^2 + \xi^3 \rangle$ is $\{\alpha^1, \alpha^2, 1 + \alpha^1 + \alpha^2\}$.

2.7.3.5. *Trace of a product*

PROPOSITION 2.34.– Let

$$x = \sum_{k=0}^{m-1} x_k e_k = \sum_{k=0}^{m-1} \tilde{x}_k \tilde{e}_k, \quad y = \sum_{k=0}^{m-1} y_k e_k = \sum_{k=0}^{m-1} \tilde{y}_k \tilde{e}_k$$

be two elements of the field $\mathbb{GF}(p^m)$ expressed in two dual bases

$$B = \{e_0, e_1, \cdots, e_{m-1}\} \text{ and } \tilde{B} = \{\tilde{e}_0, \tilde{e}_1, \cdots, \tilde{e}_{m-1}\}$$

Then, the trace of the product xy can be written as

$$\text{Tr}(xy) = \sum_{k=0}^{m-1} x_k \tilde{y}_k = \sum_{k=0}^{m-1} \tilde{x}_k y_k$$

in terms of the components of x and y in the bases B and $\tilde{B}$.

PROOF.– The proof easily follows from the linearity of the trace and Proposition 2.7.3.1. □

2.7.3.6. *Expansion coefficients*

PROPOSITION 2.35.– Let

$$x = \sum_{k=0}^{m-1} x_k e_k$$

be an element of the field $\mathbb{GF}(p^m)$ in a basis $\{e_0, e_1, \cdots, e_{m-1}\}$ with $m \geq 2$. The expansion coefficients $x_k \in \mathbb{F}_p$ (with $k = 0, 1, \cdots, m-1$) in the basis $\{e_0, e_1, \cdots, e_{m-1}\}$ are given by

$$x_k = \text{Tr}(x\tilde{e}_k), \quad k = 0, 1, \cdots, m-1$$

in terms of the elements of the dual basis $\{\tilde{e}_0, \tilde{e}_1, \cdots, \tilde{e}_{m-1}\}$ of the basis $\{e_0, e_1, \cdots, e_{m-1}\}$. Similarly,

$$x = \sum_{k=0}^{m-1} \tilde{x}_k \tilde{e}_k \Rightarrow \tilde{x}_k = \mathrm{Tr}(x e_k), \quad k = 0, 1, \cdots, m-1$$

gives the expansion coefficients in the basis $\{\tilde{e}_0, \tilde{e}_1, \cdots, \tilde{e}_{m-1}\}$.

PROOF.– It is sufficient to multiply both sides of $x = \sum_{j=0}^{m-1} x_j e_j$ by $\tilde{e}_k$ and to take the trace of both sides of the obtained relation; the use of $\mathrm{Tr}(e_j \tilde{e}_k) = \delta(j,k)$ leads to the result for x_k:

$$x\tilde{e}_k = \sum_{j=0}^{m-1} x_j e_j \tilde{e}_k$$

$$\Rightarrow \mathrm{Tr}(x\tilde{e}_k) = \sum_{j=0}^{m-1} x_j \mathrm{Tr}(e_j \tilde{e}_k) = \sum_{j=0}^{m-1} x_j \delta(j,k) = x_k$$

A similar proof yields the expression of $\tilde{x}_k$. □

2.8. Characters of a Galois field

The notion of characters is well-known for a group (see Appendix for some elements on group theory). This notion can be applied to a Galois field. Since there are two group structures for a field, it follows that there are two types of characters for a Galois field, viz. additive characters and multiplicative characters.

2.8.1. *Additive characters*

2.8.1.1. *Characters of the group* $(\mathbb{GF}(p^m), +)$

DEFINITION 2.12.– The *additive characters* $\chi_y(x)$ of the Galois field $\mathbb{GF}(p^m)$, with p prime and $m \in \mathbb{N}_1$, are defined by the applications

$$\chi_y : \mathbb{GF}(p^m) \to \mathbb{C}^*$$

$$x \mapsto \chi_y(x) = \mathrm{e}^{\mathrm{i}\frac{2\pi}{p}\mathrm{Tr}(xy)}, \quad y \in \mathbb{GF}(p^m)$$

For $y = 1$, we will abbreviate the character vector χ_1, of components $\chi_1(x)$, as χ; thus,

$$\chi = \chi_1 \Leftrightarrow \chi(x) = \mathrm{e}^{\mathrm{i}\frac{2\pi}{p}\mathrm{Tr}(x)}$$

The character vector χ is called the *canonical additive character* of $\mathbb{GF}(p^m)$. It is possible to obtain $\chi_y(x)$ from the knowledge of $\chi(xy)$ owing to

$$\chi_y(x) = \chi(xy)$$

Thus,

$$\chi_y(x) = \mathrm{e}^{\mathrm{i}\frac{2\pi}{p}xy} \Rightarrow \chi(x) = \mathrm{e}^{\mathrm{i}\frac{2\pi}{p}x}$$

when $m = 1$. For $y = 0$, the character vector χ_0, of which all components $\chi_0(x)$ are equal to 1, is called the *trivial additive character* of $\mathbb{GF}(p^m)$.

Note that $\chi_y(x) = \omega^{\mathrm{Tr}(xy)}$, where ω is any primitive p-th root of unity and $\omega^{\mathrm{Tr}(xy)}$ is well-defined since $\mathrm{Tr}(xy)$ is defined modulo p.

For fixed y and x ranging on $\mathbb{GF}(p^m)$, the characters $\chi_y(x)$ define a vector χ_y of components $\chi_y(x)$ in the space of the elements x of the group $(\mathbb{GF}(p^m), +)$. We thus have p^m character vectors χ_y in a space of dimension p^m. The additive characters of the Galois field $\mathbb{GF}(p^m)$, with p prime and $m \in \mathbb{N}_1$, are nothing but the irreducible characters of the Abelian group $(\mathbb{GF}(p^m), +)$. The p^m character vectors χ_y thus correspond to the p^m irreducible representations of the group $(\mathbb{GF}(p^m), +)$. The character vector χ_0 corresponds to the identity representation of the group $(\mathbb{GF}(p^m), +)$. In the special case where $m = 1$, the p character vectors coincide with the p one-dimensional irreducible representations of the cyclic group $(\mathbb{F}_p, +) \simeq C_p$.

2.8.1.2. *Example: additive characters of* $\mathbb{F}_3$

Table 2.28 gives all the additive characters $\chi_y(x)$ of the Galois field $\mathbb{F}_3$ in terms of powers of the primitive root of unity (of order 3)

$$\omega = \mathrm{e}^{\mathrm{i}\frac{2\pi}{3}} \Rightarrow 1 + \omega + \omega^2 = 0$$

$x \in \mathbb{F}_3 \rightarrow$ $\chi_y \downarrow$	0	1	2
χ_0	1	1	1
χ_1	1	ω	ω^2
χ_2	1	ω^2	ω

Table 2.28. *Additive characters of the Galois field $\mathbb{F}_3$: the character at the intersection of the line χ_y and the column x is $\chi_y(x) = \mathrm{e}^{\mathrm{i}\frac{2\pi}{3}xy} = \omega^{xy}$ where $\omega = \mathrm{e}^{\mathrm{i}\frac{2\pi}{3}}$*

The table is arranged in a format that is reminiscent of the one for the table of the three irreducible representations of the cyclic group C_3 isomorphic to the additive group $(\mathbb{F}_3, +)$. Note that the vector χ_2 is the complex conjugate of the vector χ_1.

2.8.1.3. *Example: additive characters of* $\mathbb{F}_5$

Table 2.29 gives all the additive characters $\chi_y(x)$ of the Galois field $\mathbb{F}_5$ in terms of powers of the primitive root of unity (of order 5)

$$\omega = \mathrm{e}^{\mathrm{i}\frac{2\pi}{5}} \Rightarrow 1 + \omega + \omega^2 + \omega^3 + \omega^4 = 0$$

$x \in \mathbb{F}_5 \rightarrow$ $\chi_y \downarrow$	0	1	2	3	4
χ_0	1	1	1	1	1
χ_1	1	ω	ω^2	ω^3	ω^4
χ_4	1	ω^4	ω^3	ω^2	ω
χ_2	1	ω^2	ω^4	ω	ω^3
χ_3	1	ω^3	ω	ω^4	ω^2

Table 2.29. *Additive characters of the Galois field $\mathbb{F}_5$: the character at the intersection of the line χ_y and the column x is $\chi_y(x) = \mathrm{e}^{\mathrm{i}\frac{2\pi}{5}xy} = \omega^{xy}$ where $\omega = \mathrm{e}^{\mathrm{i}\frac{2\pi}{5}}$*

The table is arranged in a format that is reminiscent of the one for the table of the five irreducible representations of the cyclic group C_5 isomorphic to the additive group $(\mathbb{F}_5, +)$. Note that the vector χ_4 (respectively, χ_3) is the complex conjugate of the vector χ_1 (respectively, χ_2).

The next examples deal with the additive characters of some fields $\mathbb{GF}(p^m)$ with $m = 2$ and 3.

2.8.1.4. *Example: additive characters of* $\mathbb{GF}(2^2)$

The additive characters $\chi(x)$ of $\mathbb{GF}(2^2) = \mathbb{F}_2[\xi]/\langle 1+\xi+\xi^2\rangle$ are

$$\chi(0) = +1, \quad \chi(1) = +1, \quad \chi(\alpha) = -1, \quad \chi(1+\alpha) = \chi(\alpha^2) = -1$$

where α is a root of the prime polynomial $1+\xi+\xi^2$. The complete set of additive characters of $\mathbb{GF}(2^2)$ are given in Table 2.30. This table coincides with the character table of the Klein four-group V isomorphic to $(\mathbb{GF}(2^2), +)$.

$x \in \mathbb{GF}(2^2) \rightarrow$ $\chi_y \downarrow$	0	1	α	$1+\alpha$
χ_0	1	1	1	1
χ_1	1	1	-1	-1
χ_α	1	-1	-1	1
$\chi_{1+\alpha}$	1	-1	1	-1

Table 2.30. *Additive characters of the Galois field* $\mathbb{GF}(2^2)$*: the character at the intersection of the line* χ_y *and the column* x *is* $\chi_y(x) = e^{i\frac{2\pi}{2}\mathrm{Tr}(xy)} = e^{i\pi\mathrm{Tr}(xy)}$ *(the non-zero elements of* $\mathbb{GF}(2^2)$ *are* $1 = \alpha^3$*,* α *and* $1+\alpha = \alpha^2$ *in terms of powers of the primitive element* α *root of* $1+\xi+\xi^2 = 0$*)*

2.8.1.5. *Example: additive characters of* $\mathbb{GF}(2^3)$

For $\mathbb{GF}(2^3) = \mathbb{F}_2[\xi]/\langle 1+\xi+\xi^3\rangle$, we get

$$\chi(0) = +1, \quad \chi(\alpha) = +1, \quad \chi(\alpha^2) = +1, \quad \chi(\alpha^3) = -1$$
$$\chi(\alpha^4) = +1, \quad \chi(\alpha^5) = -1, \quad \chi(\alpha^6) = -1, \quad \chi(1) = -1$$

where α is a root of $1+\xi+\xi^3$. Similarly, for $\mathbb{GF}(2^3) = \mathbb{F}_2[\xi]/\langle 1+\xi^2+\xi^3\rangle$, we obtain

$$\chi(0) = +1, \quad \chi(\alpha) = -1, \quad \chi(\alpha^2) = -1, \quad \chi(\alpha^3) = +1$$
$$\chi(\alpha^4) = -1, \quad \chi(\alpha^5) = +1, \quad \chi(\alpha^6) = +1, \quad \chi(1) = -1$$

where α is a root of $1+\xi^2+\xi^3$.

The results of this example show that the additive characters of a Galois field $\mathbb{GF}(p^m) = \mathbb{F}_p[\xi]/\langle P_m(\xi)\rangle$ depend in general on the chosen monic irreducible polynomial $P_m(\xi)$. This does not contradict the fact that two realizations of a Galois field associated with two different prime polynomials are isomorphic.

2.8.1.6. *Example: additive characters of* $\mathbb{GF}(3^2)$

Consider the field $\mathbb{GF}(3^2)$. Here, $p = 3$ and $m = 2$. We can take

$$P_2(\xi) = 2 + \xi + \xi^2$$

as a primitive polynomial in $\mathbb{F}_3[\xi]$. Then, the elements x of $\mathbb{GF}(3^2) = \mathbb{F}_3[\xi]/\langle P_2(\xi)\rangle$ are

$$x = x_0 + x_1\alpha, \quad x_0, x_1 \in \mathbb{F}_3$$

where α is a root of $P_2(\xi)$. This yields the following expressions for x

$$0, \quad 1 = \alpha^8, \quad 2 = \alpha^4, \quad \alpha, \quad 1 + \alpha = \alpha^7$$

$$2 + \alpha = \alpha^6, \quad 2\alpha = \alpha^5, \quad 1 + 2\alpha = \alpha^2, \quad 2 + 2\alpha = \alpha^3$$

Furthermore, we have

$$x = x_0 + x_1\alpha \Rightarrow \mathrm{Tr}(x) = 2x_0 + x_1\mathrm{Tr}(\alpha)$$

with

$$\mathrm{Tr}(\alpha) = \alpha + \alpha^3 = 2$$

Therefore,

$$\forall x \in \mathbb{F}_3[\xi]/\langle P_2(\xi)\rangle : \mathrm{Tr}(x) = 2(x_0 + x_1)$$

This leads to

$$\mathrm{Tr}(0) = \mathrm{Tr}(\alpha^2) = \mathrm{Tr}(\alpha^6) = 0 \quad \text{or}$$

$$\mathrm{Tr}(0) = \mathrm{Tr}(1 + 2\alpha) = \mathrm{Tr}(2 + \alpha) = 0$$

$$\mathrm{Tr}(\alpha^4) = \mathrm{Tr}(\alpha^5) = \mathrm{Tr}(\alpha^7) = 1 \quad \text{or}$$

$$\mathrm{Tr}(2) = \mathrm{Tr}(2\alpha) = \mathrm{Tr}(1 + \alpha) = 1$$

$$\mathrm{Tr}(\alpha) = \mathrm{Tr}(\alpha^3) = \mathrm{Tr}(\alpha^8) = 2 \quad \text{or}$$

$$\mathrm{Tr}(\alpha) = \mathrm{Tr}(2 + 2\alpha) = \mathrm{Tr}(1) = 2$$

Finally, the characters of $\mathbb{F}_3[\xi]/\langle P_2(\xi)\rangle$ are

$$\chi(0) = \chi(\alpha^2) = \chi(\alpha^6) = 1 \quad \text{or} \quad \chi(0) = \chi(1+2\alpha) = \chi(2+\alpha) = 1$$
$$\chi(\alpha^4) = \chi(\alpha^5) = \chi(\alpha^7) = \omega \quad \text{or} \quad \chi(2) = \chi(2\alpha) = \chi(1+\alpha) = \omega$$
$$\chi(\alpha) = \chi(\alpha^3) = \chi(\alpha^8) = \omega^2 \quad \text{or} \quad \chi(\alpha) = \chi(2+2\alpha) = \chi(1) = \omega^2$$

where

$$\omega = \mathrm{e}^{\mathrm{i}\frac{2\pi}{3}}$$

is a primitive root of unity (of order 3).

The complete set of additive characters of the field $\mathbb{F}_3[\xi]/\langle 2+\xi+\xi^2\rangle$ is given in Table 2.31. It can be proved that this table coincides with the irreducible character table of the Abelian group $C_3 \times C_3$ isomorphic to $(\mathbb{GF}(3^2), +)$. Note that $(\chi_{2+\alpha}, \chi_{1+2\alpha})$, $(\chi_{2\alpha}, \chi_\alpha)$, (χ_2, χ_1) and $(\chi_{1+\alpha}, \chi_{2+2\alpha})$ constitute four couples of complex conjugate character vectors.

$x \in \mathbb{GF}(3^2) \rightarrow$ / $\chi_y \downarrow$	0	1	2	α	$1+\alpha$	$2+\alpha$	2α	$1+2\alpha$	$2+2\alpha$
χ_0	1	1	1	1	1	1	1	1	1
$\chi_{2+\alpha}$	1	1	1	ω	ω	ω	ω^2	ω^2	ω^2
$\chi_{1+2\alpha}$	1	1	1	ω^2	ω^2	ω^2	ω	ω	ω
$\chi_{2\alpha}$	1	ω	ω^2	1	ω	ω^2	1	ω	ω^2
χ_α	1	ω^2	ω	1	ω^2	ω	1	ω^2	ω
χ_2	1	ω	ω^2	ω	ω^2	1	ω^2	1	ω
χ_1	1	ω^2	ω	ω^2	ω	1	ω	1	ω^2
$\chi_{1+\alpha}$	1	ω	ω^2	ω^2	1	ω	ω	ω^2	1
$\chi_{2+2\alpha}$	1	ω^2	ω	ω	1	ω^2	ω^2	ω	1

Table 2.31. *Additive characters of the Galois field $\mathbb{GF}(3^2) = \mathbb{F}_3[\xi]/\langle 2+\xi+\xi^2\rangle$: the character at the intersection of the line χ_y and the column x is $\chi_y(x) = \mathrm{e}^{\mathrm{i}\frac{2\pi}{3}\mathrm{Tr}(xy)} = \omega^{\mathrm{Tr}(xy)}$ where $\omega = \mathrm{e}^{\mathrm{i}\frac{2\pi}{3}}$ (the non-zero elements of $\mathbb{F}_3[\xi]/\langle 2+\xi+\xi^2\rangle$ are $1 = \alpha^8$, $2 = \alpha^4$, α, $1+\alpha = \alpha^7$, $2+\alpha = \alpha^6$, $2\alpha = \alpha^5$, $1+2\alpha = \alpha^2$ and $2+2\alpha = \alpha^3$ in terms of powers of the primitive element α root of $2+\xi+\xi^2 = 0$)*

2.8.1.7. *Properties of the additive characters*

PROPOSITION 2.36.– The additive characters of the field $\mathbb{GF}(p^m)$, p prime and m positive integer, satisfy the following properties.

– Property 1:

$$\forall x, y, z \in \mathbb{GF}(p^m) : \chi_z(x+y) = \chi_z(x)\chi_z(y)$$

– Property 2:

$$\forall y, z \in \mathbb{GF}(p^m) : \sum_{x \in \mathbb{GF}(p^m)} \chi(xy - xz) = \sum_{x \in \mathbb{GF}(p^m)} \chi_{y-z}(x)$$
$$= p^m \delta(y, z)$$

or equivalently,

$$\forall y, z \in \mathbb{GF}(p^m) : \sum_{x \in \mathbb{GF}(p^m)} \overline{\chi_z(x)}\chi_y(x) = p^m \delta(y, z)$$

– Property 3:

$$\forall x \in \mathbb{GF}(p^m) : \sum_{k=1}^{p^m-1} \chi(\alpha^k x) = p^m \delta(x, 0) - 1$$

where α is a primitive element of $\mathbb{GF}(p^m)$.

– Property 4:

$$\forall x, y \in \mathbb{GF}(p^m) : \sum_{z \in \mathbb{GF}(p^m)} \overline{\chi_z(x)}\chi_z(y) = p^m \delta(x, y)$$

to be compared with its dual relation (Property 2).

PROOF.– In Properties 2 and 4, the bar indicates complex conjugation. The proof of Properties 1 and 2 follows from properties satisfied by the irreducible representations χ_y of the Abelian group $(\mathbb{GF}(p^m), +)$. More precisely, Property 1 is a consequence of the fact that the irreducible representations of $(\mathbb{GF}(p^m), +)$ are one-dimensional; alternatively, a proof of Property 1 in the framework of the theory of Galois fields is as follows

$$\chi_z(x+y) = \mathrm{e}^{\mathrm{i}\frac{2\pi}{p}\mathrm{Tr}(xz+yz)} = \mathrm{e}^{\mathrm{i}\frac{2\pi}{p}[\mathrm{Tr}(xz)+\mathrm{Tr}(yz)]} = \mathrm{e}^{\mathrm{i}\frac{2\pi}{p}\mathrm{Tr}(xz)}\mathrm{e}^{\mathrm{i}\frac{2\pi}{p}\mathrm{Tr}(yz)}$$
$$= \chi_z(x)\chi_z(y)$$

Property 2 is a consequence of the orthogonality relation of the irreducible character vectors of the group $(\mathbb{GF}(p^m), +)$. As a corollary of Property 2, we obtain

$$\forall y \in \mathbb{GF}(p^m): \sum_{x\in\mathbb{GF}(p^m)} \chi(xy) = \sum_{x\in\mathbb{GF}(p^m)} \chi_y(x) = p^m\delta(y,0)$$

and

$$\sum_{x\in\mathbb{GF}(p^m)} \chi(x) = \sum_{x\in\mathbb{GF}(p^m)} \chi_1(x) = 0$$

(it is enough to introduce $z = 0$ and then $y = 1$ in Property 2). Property 3 follows from Property 2 by interchanging x and y and then by taking $z = 0$ and $y = 0, \alpha, \cdots, \alpha^{p^m-1}$. Finally, Property 4 trivially follows from Property 2 (Properties 2 and 4 are dual orthogonality relations, see 5.2.14). □

2.8.2. *Multiplicative characters*

2.8.2.1. *Characters of the group* $(\mathbb{GF}(p^m)^*, \times)$

DEFINITION 2.13.– The *multiplicative characters* $\psi_k(\alpha^\ell)$ of the Galois field $\mathbb{GF}(p^m)$, with p prime and m positive integer, are defined by the applications

$$\psi_k : \mathbb{GF}(p^m)^* \to \mathbb{C}$$
$$\alpha^\ell \mapsto \psi_k(\alpha^\ell) = \mathrm{e}^{\mathrm{i}\frac{2\pi}{p^m-1}k\ell}, \quad k = 0, 1, \cdots, p^m - 2$$

where α is a generator of $\mathbb{GF}(p^m)^*$ with $\ell = 0, 1, \cdots, p^m - 2$.

For fixed k, the characters $\psi_k(\alpha^\ell)$ (often abbreviated as $\psi_k(\ell)$) with α^ℓ ranging on $\mathbb{GF}(p^m)^*$ form a vector ψ_k in the space of the elements α^ℓ of the group $(\mathbb{GF}(p^m)^*, \times)$. We thus have $p^m - 1$ character vectors ψ_k in a space of dimension $p^m - 1$. The multiplicative characters of the Galois field $\mathbb{GF}(p^m)$, with p prime and m positive integer, are nothing but the irreducible characters of the group $(\mathbb{GF}(p^m)^*, \times)$ isomorphic to the cyclic group C_{p^m-1}. The $p^m - 1$ character vectors ψ_k thus correspond to the $p^m - 1$ irreducible representations of the group C_{p^m-1}. The character vector ψ_0, of which all the components $\psi_0(\alpha^\ell)$ are equal to 1, is called the *trivial multiplicative character* of $\mathbb{GF}(p^m)$; it corresponds to the identity representation of the group $(\mathbb{GF}(p^m)^*, \times)$.

The multiplicative character ψ_k of $\mathbb{GF}(p^m)$ for $k = \frac{1}{2}(p^m - 1)$ with p odd is called the *quadratic multiplicative character*. It corresponds to

$$\psi_{\frac{1}{2}(p^m-1)}(\alpha^\ell) = \mathrm{e}^{\mathrm{i}\pi\ell}$$

or equivalently,

$$\psi_{\frac{1}{2}(p^m-1)}(\alpha^\ell) = \begin{cases} +1 \text{ if } \alpha^\ell \text{ is a square in } \mathbb{GF}(p^m)^* \\ \\ -1 \text{ if } \alpha^\ell \text{ is not a square in } \mathbb{GF}(p^m)^* \end{cases}$$

where $\alpha^\ell \in \mathbb{GF}(p^m)^*$. In other words,

$$\psi_{\frac{1}{2}(p^m-1)}(x) = \left(\frac{x}{p}\right), \quad x \in \mathbb{GF}(p^m)^*$$

where $(\frac{x}{p})$ is a Legendre symbol (see 5.1.6 in Chapter 5). The term *quadratic* comes from the fact that the order of the character vector ψ_k with $k = \frac{1}{2}(p^m - 1)$ is equal to 2 (i.e. for each x in $\mathbb{GF}(p^m)^*$, we have $\psi_k(x)^2 = 1$). For $m = 1$, we have

$$\psi_{\frac{1}{2}(p-1)}(a) = \left(\frac{a}{p}\right) \equiv a^{\frac{p-1}{2}} \bmod p, \quad a \in \mathbb{F}_p{}^*$$

as a particular case.

Note that $\psi_k(x)$ with $k = 0, 1, \cdots, p^m - 2$ are not defined for $x = 0$. It is useful to extend the definition of the character vector ψ_k by taking $\psi_k(0) = 0$ for $k \neq 0$ and $\psi_0(0) = 1$ for $k = 0$.

In the special case where $m = 1$, the $p - 1$ character vectors coincide with the p one-dimensional irreducible representations of the cyclic group C_{p-1}.

2.8.2.2. *Example: multiplicative characters of* $\mathbb{F}_3$

Table 2.32 gives all the multiplicative characters $\psi_k(\ell)$ of the Galois field $\mathbb{F}_3$. This table coincides with the character table of the cyclic group C_2 (the group $(\mathbb{F}_3{}^*, \times)$ is isomorphic to C_2).

$\alpha^\ell \in \mathbb{F}_3{}^* \rightarrow$ $\psi_k \downarrow$	$\alpha^0 = 1$	$\alpha^1 = 2$
ψ_0	1	1
ψ_1	1	-1

Table 2.32. *Multiplicative characters of the Galois field* $\mathbb{F}_3$*: the character at the intersection of the line* ψ_k *and the column* α^ℓ *is* $\psi_k(\alpha^\ell) = \mathrm{e}^{\mathrm{i}\frac{2\pi}{2}k\ell} = (-1)^{k\ell}$ *where* α *(*$= \alpha^1 = 2$*) is a primitive element of* $\mathbb{F}_3$

2.8.2.3. *Example: multiplicative characters of* $\mathbb{F}_5$

Table 2.33 gives all the multiplicative characters $\psi_k(\ell)$ of the Galois field $\mathbb{F}_5$. This table coincides with the character table of the cyclic group C_4 (the group $(\mathbb{F}_5{}^*, \times)$ is isomorphic to C_4).

$\alpha^\ell \in \mathbb{F}_5{}^* \rightarrow$ $\psi_k \downarrow$	$\alpha^0 = 1$	$\alpha^1 = 2$	$\alpha^2 = 4$	$\alpha^3 = 3$
ψ_0	1	1	1	1
ψ_1	1	i	-1	$-$i
ψ_2	1	-1	1	-1
ψ_3	1	$-$i	-1	i

Table 2.33. *Multiplicative characters of the Galois field* $\mathbb{F}_5$*: the character at the intersection of the line* ψ_k *and the column* α^ℓ *is* $\psi_k(\alpha^\ell) = \mathrm{e}^{\mathrm{i}\frac{2\pi}{4}k\ell} = \mathrm{i}^{k\ell}$ *where* α *(*$= \alpha^1 = 2$*) is a primitive element of* $\mathbb{F}_5$

2.8.2.4. *Example: multiplicative characters of* $\mathbb{GF}(2^2)$

Table 2.34 gives all the multiplicative characters $\psi_k(\ell)$ of the Galois field $\mathbb{GF}(2^2)$. This table coincides with the character table of the cyclic group C_3 (the group $(\mathbb{GF}(2^2)^*, \times)$ is isomorphic to C_3). The entries of Table 2.34 are the same as those of Table 2.28 because the groups $(\mathbb{GF}(2^2)^*, \times)$ and $(\mathbb{F}_3, +)$ are isomorphic.

$\alpha^\ell \in \mathbb{GF}(2^2)^* \rightarrow$ $\psi_k \downarrow$	α^0	α^1	α^2
ψ_0	1	1	1
ψ_1	1	ω	ω^2
ψ_2	1	ω^2	ω

Table 2.34. *Multiplicative characters of the Galois field $\mathbb{GF}(2^2)$: the character at the intersection of the line ψ_k and the column α^ℓ is $\psi_k(\alpha^\ell) = e^{i\frac{2\pi}{3}k\ell} = \omega^{k\ell}$ where $\omega = e^{i\frac{2\pi}{3}}$ and α ($= \alpha^1$) is a primitive element of $\mathbb{GF}(2^2)$*

2.8.2.5. *Example: multiplicative characters of* $\mathbb{GF}(2^3)$

Table 2.35 gives all the multiplicative characters $\psi_k(\alpha^\ell)$ of the Galois field $\mathbb{GF}(2^3)$ in terms of powers of the primitive root of unity (of order 7)

$$\omega = e^{i\frac{2\pi}{7}} \Rightarrow 1 + \omega + \omega^2 + \cdots + \omega^6 = 0$$

The table is arranged in a format that is reminiscent of the one for the table of the seven irreducible representations of the cyclic group C_7 isomorphic to the multiplicative group $(\mathbb{GF}(2^3)^*, \times)$.

$\alpha^\ell \in \mathbb{GF}(2^3)^* \rightarrow$ $\psi_k \downarrow$	α^0	α^1	α^2	α^3	α^4	α^5	α^6
ψ_0	1	1	1	1	1	1	1
ψ_1	1	ω	ω^2	ω^3	ω^4	ω^5	ω^6
ψ_2	1	ω^2	ω^4	ω^6	ω	ω^3	ω^5
ψ_3	1	ω^3	ω^6	ω^2	ω^5	ω	ω^4
ψ_4	1	ω^4	ω	ω^5	ω^2	ω^6	ω^3
ψ_5	1	ω^5	ω^3	ω	ω^6	ω^4	ω^2
ψ_6	1	ω^6	ω^5	ω^4	ω^3	ω^2	ω

Table 2.35. *Multiplicative characters of the Galois field $\mathbb{GF}(2^3)$: the character at the intersection of the line ψ_k and the column α^ℓ is $\psi_k(\alpha^\ell) = e^{i\frac{2\pi}{7}k\ell} = \omega^{k\ell}$ where $\omega = e^{i\frac{2\pi}{7}}$ and α ($= \alpha^1$) is a primitive element of $\mathbb{GF}(2^3)$*

2.8.2.6. *Example: multiplicative characters of* $\mathbb{GF}(3^2)$

Table 2.36 gives all the multiplicative characters $\psi_k(\alpha^\ell)$ of the Galois field $\mathbb{GF}(3^2)$ in terms of powers of the primitive root of unity (of order 8)

$$\omega = e^{i\frac{2\pi}{8}} = e^{i\frac{\pi}{4}} \Rightarrow 1 + \omega + \omega^2 + \cdots + \omega^7 = 0$$

The table is arranged in a format that is reminiscent of the one for the table of the eight irreducible representations of the cyclic group C_8 isomorphic to the multiplicative group $(\mathbb{GF}(3^2)^*, \times)$.

$\alpha^\ell \in \mathbb{GF}(3^2)^* \rightarrow$ $\psi_k \downarrow$	α^0	α^1	α^2	α^3	α^4	α^5	α^6	α^7
ψ_0	1	1	1	1	1	1	1	1
ψ_1	1	ω	i	iω	-1	$-\omega$	$-$i	$-$iω
ψ_2	1	i	-1	$-$i	1	i	-1	$-$i
ψ_3	1	iω	$-$i	ω	-1	$-$iω	i	$-\omega$
ψ_4	1	-1	1	-1	1	-1	1	-1
ψ_5	1	$-\omega$	i	$-$iω	-1	ω	$-$i	iω
ψ_6	1	$-$i	-1	i	1	$-$i	-1	i
ψ_7	1	$-$iω	$-$i	$-\omega$	-1	iω	i	ω

Table 2.36. *Multiplicative characters of the Galois field* $\mathbb{GF}(3^2)$*: the character at the intersection of the line* ψ_k *and the column* α^ℓ *is* $\psi_k(\alpha^\ell) = \mathrm{e}^{\mathrm{i}\frac{2\pi}{8}k\ell} = \omega^{k\ell}$ *where* $\omega = \mathrm{e}^{\frac{\mathrm{i}\pi}{4}}$ *and* α $(= \alpha^1)$ *is a primitive element of* $\mathbb{GF}(3^2)$

2.8.2.7. *Properties of the multiplicative characters*

PROPOSITION 2.37.– Let α be a primitive element of the field $\mathbb{GF}(p^m)$, p prime and m positive integer. The multiplicative characters of $\mathbb{GF}(p^m)$ satisfy the following properties:

– Property 1:

$$\forall \alpha^\ell, \alpha^{\ell'} \in \mathbb{GF}(p^m)^* : \psi_k(\alpha^{\ell+\ell'}) = \psi_k(\alpha^\ell)\psi_k(\alpha^{\ell'})$$

where $k = 0, 1, \cdots, p^m - 2$ and $\ell + \ell'$ is defined modulo $p^m - 1$.

– Property 2:

$$\forall k, k' \in \{0, 1, \cdots, p^m - 2\} : \sum_{\ell=0}^{p^m-2} \overline{\psi_k(\alpha^\ell)}\psi_{k'}(\alpha^\ell) = (p^m - 1)\delta(k, k')$$

– Property 3:

$$\forall \ell, \ell' \in \{0, 1, \cdots, p^m - 2\} : \sum_{k=0}^{p^m-2} \overline{\psi_k(\alpha^\ell)}\psi_k(\alpha^{\ell'}) = (p^m - 1)\delta(\ell, \ell')$$

to be compared with its dual relation (Property 2).

PROOF.– The proofs of Properties 1, 2 and 3 follow from properties satisfied by the irreducible representations of the cyclic group C_{p^m-1}: Property 1 is trivial and Properties 2 and 3 reflect the orthogonality relations of the irreducible character vectors of the group C_{p^m-1}. □

2.8.2.8. *Remark*

Interestingly, note that Properties 1 and 2 of multiplicative and additive characters are comparable to

$$\omega^{m+n} = \omega^m \omega^n \text{ and } \sum_{n=0}^{d-1} \omega^{n(m-\ell)} = d\delta(m, \ell)$$

where $\omega = \mathrm{e}^{\mathrm{i}\frac{2\pi}{d}}$ is a primitive d-th root of unity, respectively.

2.9. Gaussian sums over Galois fields

2.9.1. *Gauss sum over* $\mathbb{Z}_d$

The quadratic sum

$$G(d) = \sum_{n \in \mathbb{Z}_d} \mathrm{e}^{\mathrm{i}\frac{2\pi}{d}n^2}, \quad d \in \mathbb{N}_1$$

is called a Gauss sum. The evaluation of $G(d)$ was achieved by Gauss (see section 5.1.7.1 of Chapter 5). In the case where the ring $\mathbb{Z}_d$ is replaced by the field $\mathbb{F}_p$, we have

$$G(p) = \sum_{n \in \mathbb{F}_p} \mathrm{e}^{\mathrm{i}\frac{2\pi}{p}n^2}, \quad p \text{ (even or odd) prime}$$

Furthermore,

$$|G(p)| = \sqrt{p}$$

for p odd prime.

2.9.2. *Quadratic Gauss sum and quadratic characters*

The quadratic Gauss sum $G(p)$ with p odd prime can be rewritten as

$$\sum_{n\in\mathbb{F}_p} e^{i\frac{2\pi}{p}n^2} = \sum_{n=0}^{p-1}\left(\frac{n}{p}\right)e^{i\frac{2\pi}{p}n} = \sum_{n=1}^{p-1}\left(\frac{n}{p}\right)e^{i\frac{2\pi}{p}n}$$

in terms of Legendre symbols. Therefore,

$$\sum_{n\in\mathbb{F}_p} e^{i\frac{2\pi}{p}n^2} = \sum_{n=0}^{p-1}\psi_{\frac{1}{2}(p-1)}(n)e^{i\frac{2\pi}{p}n} = \sum_{n=1}^{p-1}\psi_{\frac{1}{2}(p-1)}(n)e^{i\frac{2\pi}{p}n}$$

where $\psi_{\frac{1}{2}(p-1)}$ is the quadratic multiplicative character of $\mathbb{F}_p$.

2.9.2.1. *Example:* $p = 3$

We have

$$\sum_{n\in\mathbb{F}_3} e^{i\frac{2\pi}{3}n^2} = 1 + \omega + \omega^4 = 1 + 2\omega$$

and

$$\sum_{n=1}^{2}\left(\frac{n}{3}\right)e^{i\frac{2\pi}{3}n} = \omega - \omega^2 = 1 + 2\omega$$

$$\sum_{n=1}^{2}\psi_1(n)e^{i\frac{2\pi}{3}n} = \omega - \omega^2 = 1 + 2\omega$$

where $\omega = e^{i\frac{2\pi}{3}}$.

2.9.2.2. *Example:* $p = 5$

As a more elaborate example, it can be checked that

$$\sum_{n\in\mathbb{F}_5} e^{i\frac{2\pi}{5}n^2} = \sum_{n=1}^{4}\left(\frac{n}{5}\right)e^{i\frac{2\pi}{5}n} = \sum_{n=1}^{4}\psi_2(n)e^{i\frac{2\pi}{5}n} = 1 + 2\omega + 2\omega^4$$

where $\omega = e^{i\frac{2\pi}{5}}$.

2.9.3. *Gauss sum over* $\mathbb{GF}(p^m)$

Let us consider the Gaussian sum

$$G_m(\psi_k, \chi_y) = \sum_{x \in \mathbb{GF}(p^m)^*} \psi_k(x)\chi_y(x)$$

where ψ_k is an arbitrary multiplicative character vector of $(\mathbb{GF}(p^m)^*, \times)$ and χ_y an arbitrary additive character vector of $(\mathbb{GF}(p^m), +)$, and the summation on x runs over all the units of $\mathbb{GF}(p^m)$. Interestingly, for $m = 1$, it should be observed that $G_1(\psi_k, \chi_y)$ looks like a discrete Fourier transform of the function ψ_k. Note that

$$G_1\left(\psi_{\frac{1}{2}(p-1)}, \chi_1\right) = \sum_{n \in \mathbb{F}_p{}^*} \left(\frac{n}{p}\right) \mathrm{e}^{\mathrm{i}\frac{2\pi}{p}n} = G(p)$$

(see 2.9.2). Therefore, the Gaussian sum $G_m(\psi_k, \chi_y)$ is a generalization of the Gauss sum $G(p)$.

It is easily checked that

$$\begin{aligned}
G_m(\psi_0, \chi_0) &= p^m - 1 \\
G_m(\psi_0, \chi_y) &= -1 \text{ for } \chi_y \neq \chi_0 \\
G_m(\psi_k, \chi_0) &= 0 \text{ for } \psi_k \neq \psi_0 \\
|G_m(\psi_k, \chi_y)| &= \sqrt{p^m} \text{ for } \psi_k \neq \psi_0,\ \chi_y \neq \chi_0
\end{aligned}$$

where ψ_0 is the trivial multiplicative character vector of $(\mathbb{GF}(p^m)^*, \times)$ and χ_0 the trivial additive character vector of $(\mathbb{GF}(p^m), +)$. By using the extended definition of the character vector ψ_k (viz. $\psi_k(0) = 0$ for $k \neq 0$ and $\psi_0(0) = 1$), we can write

$$\psi_k(0)\chi_y(0) + G_m(\psi_k, \chi_y) = \sum_{x \in \mathbb{GF}(p^m)} \psi_k(x)\chi_y(x)$$

where the summation on x runs on all the elements of $\mathbb{GF}(p^m)$. We are thus led to

$$\sum_{x\in\mathbb{GF}(p^m)} \psi_0(x)\chi_0(x) = p^m$$

$$\sum_{x\in\mathbb{GF}(p^m)} \psi_0(x)\chi_y(x) = 0 \text{ for } \chi_y \neq \chi_0$$

$$\sum_{x\in\mathbb{GF}(p^m)} \psi_k(x)\chi_0(x) = 0 \text{ for } \psi_k \neq \psi_0$$

$$\left|\sum_{x\in\mathbb{GF}(p^m)} \psi_k(x)\chi_y(x)\right| = \sqrt{p^m} \text{ for } \psi_k \neq \psi_0,\ \chi_y \neq \chi_0$$

Only the first two relations are affected by replacing the sum over $x \in \mathbb{GF}(p^m)^*$ with the sum over $x \in \mathbb{GF}(p^m)$.

2.9.4. *Weil sum over* $\mathbb{GF}(p^m)$

Let χ_y be a non-trivial additive character of the Galois field $\mathbb{GF}(p^m)$ and

$$f_n(\xi) = a_n\xi^n + a_{n-1}\xi^{n-1} + \cdots + a_1\xi$$

be a polynomial of positive degree n in $\mathbb{GF}(p^m)[\xi]$. The sum

$$\sum_{x\in\mathbb{GF}(p^m)} \chi_y(f_n(x)), \quad p \text{ odd}, \quad n < p^m$$

is called a Weil sum. We have (Weil's theorem)

$$\left|\sum_{x\in\mathbb{GF}(p^m)} \chi_y(f_n(x))\right| \leq (n-1)\sqrt{p^m}$$

for $\gcd(a_n, p^m) = 1$. In the particular case where $n = 2$, we get

$$\left|\sum_{x\in\mathbb{GF}(p^m)} \chi_y(f_2(x))\right| = \sqrt{p^m}$$

for p odd prime and $y \neq 0$.

An important relation (used in Chapter 4) is

$$\left| \sum_{x \in \mathbb{GF}(p^m)} \chi_1(ax^2 + bx) \right| = \sqrt{p^m}$$

or in detailed form

$$\left| \sum_{x \in \mathbb{GF}(p^m)} \mathrm{e}^{\mathrm{i}\frac{2\pi}{p}\mathrm{Tr}(ax^2+bx)} \right| = \sqrt{p^m}$$

for p odd prime, $a \in \mathbb{GF}(p^m)^*$ and $b \in \mathbb{GF}(p^m)$. In the special case where $m = 1$, we have

$$\left| \sum_{n \in \mathbb{F}_p} \mathrm{e}^{\mathrm{i}\frac{2\pi}{p}(an^2+bn)} \right| = \sqrt{p}$$

for p odd prime, $(a, p) = 1$ and $b \in \mathbb{Z}$ (see also 5.1.7.1).

3

Galois Rings

Galois rings are special finite rings. They play an important role in the theory of finite rings. As for the Galois fields, there exists a polynomial construction for Galois rings. Indeed, a construction of an arbitrary Galois ring can be achieved via a Galois extension of a base ring. Such a construction is very similar to that of a Galois field from a base field.

Galois rings are useful in classical information, especially in coding theory (in particular for linear codes). They are also of interest in quantum information for the construction of mutually unbiased bases. Along the vein of Galois fields, they open new perspectives in quantum information at the level of quantum logic, intrication of quantum states, quantum cryptography and quantum error corrections.

For the present chapter, the reader should consult the references in the sections *Mathematical literature: rings and fields* and *Useful web links* of the bibliography, in particular, the books by B. R. McDonald and Z.-X. Wan give a detailed presentation of Galois rings. The sections *Mathematical literature: number theory* and *Theoretical physics literature: MUBs* contain useful references concerning Gaussian sums over Galois rings.

3.1. Generalities

3.1.1. *Principal ideal of a commutative ring*

DEFINITION 3.1.– A non-empty subset I of a commutative ring $(R, +, \times)$ such that

$$\forall r \in R, \ \forall i \in I, \ \forall j \in I : i + j \in I, \ i \times r \in I$$

is called an ideal of $(R, +, \times)$. An ideal I of a commutative ring $(R, +, \times)$ is said to be principal if it is generated by a single element a of R: $I = aR = \{a \times r \mid r \in R\}$. The principal ideal generated by a is denoted as $\langle a \rangle$.

3.1.2. *Galois ring*

3.1.2.1. *Defining a Galois ring*

DEFINITION 3.2.– A finite unitary (commutative) ring R such that the set of its zero divisors including 0 constitutes a principal ideal $\langle p \rangle$ with p prime (i.e. $R/\langle p \rangle$ is an integrity ring) is called a *Galois ring*.

As a matter of fact, the principal ideal spanned by the zero divisors (including the trivial zero divisor) of R is the sole maximal ideal of R.

3.1.2.2. *Counter-example:* $\mathbb{Z}_6$

The ring $\mathbb{Z}_6$ is a finite (commutative) ring with unity for which the set of zero divisors is $\{0, 2, 3, 4\}$. This set does not constitute an ideal of $\mathbb{Z}_6$. Consequently, $\mathbb{Z}_6$ is not a Galois ring.

3.1.2.3. *Example:* $\mathbb{Z}_4$, $\mathbb{Z}_8$ *and* $\mathbb{Z}_9$

The rings $\mathbb{Z}_4$, $\mathbb{Z}_8$ and $\mathbb{Z}_9$ are finite (commutative) rings with unity for which the set of zero divisors (including 0) constitutes a principal ideal $\langle p = 2 \rangle$, $\langle p = 2 \rangle$ and $\langle p = 3 \rangle$, respectively (see Table 3.1). Therefore, $\mathbb{Z}_4 (= \mathbb{Z}_{2^2})$, $\mathbb{Z}_8 (= \mathbb{Z}_{2^3})$ and $\mathbb{Z}_9 (= \mathbb{Z}_{3^2})$ are Galois rings. Note that $\langle p = 2 \rangle$, $\langle p = 2 \rangle$ and $\langle p = 3 \rangle$ are the unique maximal ideals of $\mathbb{Z}_{2^2}$, $\mathbb{Z}_{2^3}$ and $\mathbb{Z}_{3^2}$,respectively.

	Set of Zero Divisors	Principal Ideal $\langle p \rangle$
$\mathbb{Z}_{2^2}$	$\{0, 2\}$	$\langle 2 \rangle$
$\mathbb{Z}_{2^3}$	$\{0, 2, 4, 6\}$	$\langle 2 \rangle$
$\mathbb{Z}_{3^2}$	$\{0, 3, 6\}$	$\langle 3 \rangle$

Table 3.1. *Set of zero divisors spanning a principal ideal for the rings* $\mathbb{Z}_{2^2}$, $\mathbb{Z}_{2^3}$ *and* $\mathbb{Z}_{3^2}$

3.1.2.4. *Example:* $\mathbb{Z}_{p^s}$

More generally, the ring $\mathbb{Z}_{p^s}$ of integers modulo p^s with p prime and s positive integer is a Galois ring (1 is the identity of $\mathbb{Z}_{p^s}$; the zero divisors including 0 of $\mathbb{Z}_{p^s}$ form the principal ideal $\langle p \rangle$ of the finite ring $\mathbb{Z}_{p^s}$; indeed, $\langle p \rangle$ is the unique maximal ideal of $\mathbb{Z}_{p^s}$ and the ring $\mathbb{Z}_{p^s}/\langle p \rangle$ is isomorphic to the field $\mathbb{F}_p$). The Galois ring $\mathbb{Z}_{p^s}$ has p^s elements and is of characteristic p^s.

In the special case $s = 1$, we have $\mathbb{Z}_{p^1} = \mathbb{F}_p$ for which the only zero divisor is the trivial zero divisor 0. The zero ideal $\langle 0 \rangle$ of $\mathbb{Z}_p$ is a principal ideal (here, $\langle p \rangle = \langle 0 \rangle$). Consequently, $\mathbb{Z}_p$ is a Galois ring. The Galois ring $\mathbb{Z}_p$ is nothing but the Galois field $\mathbb{F}_p$.

3.2. Construction of a Galois ring

3.2.1. *Elements of* $\mathbb{Z}_{p^s}$

3.2.1.1. *Writing an element of* $\mathbb{Z}_{p^s}$

PROPOSITION 3.1.– Every element a of the ring $\mathbb{Z}_{p^s}$ with p prime number and s positive integer can be developed as

$$a = d_0 \times 1 + d_1 \times p + \cdots + d_{s-1} \times p^{s-1}$$

or simply as

$$a = d_0 + d_1 p + \cdots + d_{s-1} p^{s-1}$$

where each coefficient d_i belongs to the field $\mathbb{F}_p$ ($i = 0, 1, \cdots, s-1$) and where the addition $+$ and the multiplication $\times$ act in the ring $\mathbb{Z}_{p^s}$ (the sign $\times$ in the development of a is often omitted).

This result is trivial for $s = 1$ that corresponds to the field $\mathbb{F}_p$.

3.2.1.2. *Example:* $\mathbb{Z}_{2^2}$ *and* $\mathbb{Z}_{2^3}$

For $s = 2$ and $p = 2$, we readily verify that

$$\forall a \in \mathbb{Z}_{2^2} : a = d_0 + 2d_1, \quad d_i \in \mathbb{F}_2 \ (i = 0, 1)$$

so that the elements a, denoted as (d_0, d_1), of $\mathbb{Z}_{2^2}$ are

$$0 = (0,0), \quad 1 = (1,0), \quad 2 = (0,1), \quad 3 = (1,1)$$

Similarly,

$$\forall a \in \mathbb{Z}_{2^3} : a = d_0 + 2d_1 + 2^2 d_2, \quad d_i \in \mathbb{F}_2 \ (i = 0, 1, 2)$$

and the elements a, denoted as (d_0, d_1, d_2), of $\mathbb{Z}_{2^3}$ are

$$\begin{aligned} &0 = (0,0,0), \quad 1 = (1,0,0), \quad 2 = (0,1,0), \quad 3 = (1,1,0) \\ &4 = (0,0,1), \quad 5 = (1,0,1), \quad 6 = (0,1,1), \quad 7 = (1,1,1) \end{aligned}$$

3.2.1.3. *Example:* $\mathbb{Z}_{3^2}$

In much the same way, we have

$$\forall a \in \mathbb{Z}_{3^2} : a = d_0 + 3d_1, \quad d_i \in \mathbb{F}_3 \ (i = 0, 1)$$

This leads to the following elements a, denoted as (d_0, d_1), of $\mathbb{Z}_{3^2}$

$$\begin{aligned} &0 = (0,0), \quad 1 = (1,0), \quad 2 = (2,0), \quad 3 = (0,1) \\ &4 = (1,1), \quad 5 = (2,1), \quad 6 = (0,2), \quad 7 = (1,2), \quad 8 = (2,2) \end{aligned}$$

3.2.2. *The* $\mathbb{Z}_{p^s} \to \mathbb{Z}_p$ *and* $\mathbb{Z}_{p^s}[\xi] \to \mathbb{Z}_p[\xi]$ *homomorphisms*

3.2.2.1. *Two ring homomorphisms*

PROPOSITION 3.2.– For p prime number and s positive integer, let $\mathbb{Z}_{p^s} \to \mathbb{Z}_p$ be the map defined by

$$a = d_0 + d_1 p + \cdots + d_{s-1} p^{s-1} \in \mathbb{Z}_{p^s} \mapsto \overline{a} = d_0 \in \mathbb{Z}_p$$

where d_i belongs to $\mathbb{Z}_p$ for $i = 0, 1, \cdots, s-1$ and let $\mathbb{Z}_{p^s}[\xi] \to \mathbb{Z}_p[\xi]$ be the map defined by

$$f(\xi) = c_0 + c_1\xi + \cdots + c_n\xi^n \in \mathbb{Z}_{p^s}[\xi]$$
$$\mapsto \overline{f(\xi)} = \overline{c_0} + \overline{c_1}\xi + \cdots + \overline{c_n}\xi^n \in \mathbb{Z}_p[\xi]$$

where c_j and $\overline{c_j}$ belong to $\mathbb{Z}_{p^s}$ and $\mathbb{Z}_p$, respectively, for $j = 0, 1, \cdots, n$ (ξ is an indeterminate over $\mathbb{Z}_{p^s}$ and $\mathbb{Z}_p$). The maps

$$\mathbb{Z}_{p^s} \to \mathbb{Z}_p = \mathbb{F}_p$$
$$a \mapsto \overline{a}$$

and

$$\mathbb{Z}_{p^s}[\xi] \to \mathbb{Z}_p[\xi] = \mathbb{F}_p[\xi]$$
$$f(\xi) \mapsto \overline{f(\xi)}$$

are two ring homomorphisms.

PROOF.– For the map $\mathbb{Z}_{p^s} \to \mathbb{Z}_p$, it is easy to show that

$$\forall a, b \in \mathbb{Z}_{p^s} : \overline{a+b} = \overline{a} + \overline{b}, \quad \overline{a \times b} = \overline{a} \times \overline{b}$$

This result can be used in turn to prove that

$$\forall f(\xi), g(\xi) \in \mathbb{Z}_{p^s}[\xi] : \overline{f(\xi) + g(\xi)} = \overline{f(\xi)} + \overline{g(\xi)}$$
$$\overline{f(\xi)\, g(\xi)} = \overline{f(\xi)}\, \overline{g(\xi)}$$

for the map $\mathbb{Z}_{p^s}[\xi] \to \mathbb{Z}_p[\xi]$, with evident notations. □

Note that $\overline{a} = a \bmod p$ ($\overline{a}$ is the residue of a modulo p) and $\overline{f(\xi)} = f(\xi) \bmod p$.

For $s = 1$, the maps $\mathbb{Z}_{p^1} \to \mathbb{Z}_p$ and $\mathbb{Z}_{p^1}[\xi] \to \mathbb{Z}_p[\xi]$ are identity maps.

3.2.2.2. *Example:* $\mathbb{Z}_{2^2} \to \mathbb{Z}_2$

We have

$$\overline{0} = 0, \quad \overline{1} = 1, \quad \overline{2} = 0, \quad \overline{3} = 1$$

that follows from the correspondence $d_0 + 2d_1 \in \mathbb{Z}_{2^2} \mapsto d_0 \in \mathbb{Z}_2$.

3.2.2.3. *Example:* $\mathbb{Z}_{2^2}[\xi] \to \mathbb{Z}_2[\xi]$

We have

– $f(\xi) = 3 + \xi \in \mathbb{Z}_{2^2}[\xi] \mapsto \overline{f(\xi)} = 1 + \xi \in \mathbb{Z}_2[\xi]$

– $f(\xi) = 1 + 3\xi \in \mathbb{Z}_{2^2}[\xi] \mapsto \overline{f(\xi)} = 1 + \xi \in \mathbb{Z}_2[\xi]$

– $f(\xi) = 1 + 2\xi + 3\xi^2 \in \mathbb{Z}_{2^2}[\xi] \mapsto \overline{f(\xi)} = 1 + \xi^2 \in \mathbb{Z}_2[\xi]$

that follows from the correspondences $c_0 + c_1\xi \in \mathbb{Z}_{2^2}[\xi] \mapsto \overline{c_0} + \overline{c_1}\xi \in \mathbb{Z}_2[\xi]$ and $c_0 + c_1\xi + c_2\xi^2 \in \mathbb{Z}_{2^2}[\xi] \mapsto \overline{c_0} + \overline{c_1}\xi + \overline{c_2}\xi^2 \in \mathbb{Z}_2[\xi]$.

3.2.3. *Basic irreducible polynomial*

3.2.3.1. *Monic basic irreducible (primitive) polynomial*

DEFINITION 3.3.– Let $P_m(\xi)$ be a monic polynomial of degree m in $\mathbb{Z}_{p^s}[\xi]$ (p prime and m positive integer). If its image $\overline{P_m(\xi)}$ in $\mathbb{Z}_p[\xi]$ is irreducible over $\mathbb{Z}_p$, then $P_m(\xi)$ is called a *monic basic irreducible polynomial* over $\mathbb{Z}_{p^s}$.

In the preceding definition, the word *irreducible* can be replaced twice by *primitive*. Every *monic basic primitive polynomial* is a monic basic irreducible polynomial, but the reverse is not true in general.

It can be shown that, for any positive integer m, there exist monic basic primitive (and therefore irreducible) polynomials $P_m(\xi)$ of degree m over $\mathbb{Z}_{p^s}$ that divide (modulo p^s) $\xi^{p^m-1} - 1$ in $\mathbb{Z}_{p^s}[\xi]$.

Note that $\overline{P_m(\xi)} = P_m(\xi)$ modulo p and that for $s = 1$, the polynomial $\overline{P_m(\xi)}$ coincides with $P_m(\xi)$.

3.2.3.2. *Example: monic basic irreducible polynomials in* $\mathbb{Z}_{2^2}[\xi]$

We have

– $P_1(\xi) = 3 + \xi \in \mathbb{Z}_{2^2}[\xi] \mapsto \overline{P_1(\xi)} = 1 + \xi \in \mathbb{Z}_2[\xi]$

– $P_2(\xi) = 1 + \xi + \xi^2 \in \mathbb{Z}_{2^2}[\xi] \mapsto \overline{P_2(\xi)} = 1 + \xi + \xi^2 \in \mathbb{Z}_2[\xi]$

– $P_3(\xi) = 3 + \xi + 2\xi^2 + \xi^3 \in \mathbb{Z}_{2^2}[\xi] \mapsto \overline{P_3(\xi)} = 1 + \xi + \xi^3 \in \mathbb{Z}_2[\xi]$

– $P_4(\xi) = 1 + 3\xi + 2\xi^2 + \xi^4 \in \mathbb{Z}_{2^2}[\xi] \mapsto \overline{P_4(\xi)} = 1 + \xi + \xi^4 \in \mathbb{Z}_2[\xi]$

Therefore, each of the polynomials $P_m(\xi)$, $m = 1$ to 4, is a monic basic irreducible polynomial over $\mathbb{Z}_{2^2}$. Observe that $\overline{P_m(\xi)} = P_m(\xi)$ modulo $p = 2$ and that the polynomial $P_m(\xi)$ divides (modulo $p^s = 2^2$) $\xi^{2^m-1} - 1$ in $\mathbb{Z}_{2^2}[\xi]$, $m = 1$ to 4.

3.2.4. *Extension of a base ring*

PROPOSITION 3.3.– Let $P_m(\xi)$ be a monic basic primitive polynomial of degree m over the Galois ring $\mathbb{Z}_{p^s}$ (with m and s positive integers, p prime number). Then, the residue class ring

$$\mathbb{Z}_{p^s}[\xi]/\langle P_m(\xi)\rangle$$

is a Galois ring denoted as $\mathbb{GR}(p^s, m)$. This ring is of characteristic p^s and of cardinal $(p^s)^m = p^{sm}$. Any element a of $\mathbb{GR}(p^s, m)$ can be written as

$$a = a_0 + a_1\alpha + \cdots + a_{m-1}\alpha^{m-1}, \quad a_i \in \mathbb{Z}_{p^s}\ (i = 0, 1, \cdots, m-1)$$

where α is a non-zero element of order $p^m - 1$ (i.e. $\alpha^{p^m-1} = 1$) which is a root of $P_m(\xi)$, with $P_m(\xi)$ dividing $\xi^{p^m-1} - 1$ in $\mathbb{Z}_{p^s}[\xi]$.

ELEMENT OF PROOF.– The zero divisors of $\mathbb{GR}(p^s, m)$, including the trivial zero divisor 0, constitute the principal ideal $\langle p\rangle$ of $\mathbb{GR}(p^s, m)$ (in fact, the sole maximal ideal of $\mathbb{GR}(p^s, m)$). □

The Galois ring $\mathbb{GR}(p^s, m)$ is said to be a Galois extension of degree m of the ring $\mathbb{Z}_{p^s}$ of characteristic p^s. The ring $\mathbb{Z}_{p^s}$ is called a *base ring* or *prime ring*. The Galois ring $\mathbb{GR}(p^s, m)$ is the unique (up to isomorphism) extension of degree m of the ring $\mathbb{Z}_{p^s}$ of integers modulo p^s.

The structure of a Galois ring is characterized by its characteristic

$$\mathrm{charact}(\mathbb{GR}(p^s, m)) = \mathrm{charact}(\mathbb{Z}_{p^s}) = p^s$$

that is a positive power p^s of a prime number p and by its cardinal

$$\mathrm{Card}(\mathbb{GR}(p^s, m)) = p^{sm}$$

that is a positive power $(p^s)^m = p^{sm}$ of the characteristic p^s ($s \geq 1$ and $m \geq 1$).

Note that the two ring homomorphisms defined in 3.2.2.1 induce a new ring homomorphism, namely,

$$\mathbb{Z}_{p^s}[\xi]/\langle P_m(\xi)\rangle = \mathbb{GR}(p^s, m) \to \mathbb{F}_p[\xi]/\langle\overline{P_m(\xi)}\rangle = \mathbb{GF}(p^m)$$
$$a_0 + a_1\alpha + ... + a_{m-1}\alpha^{m-1} \mapsto \overline{a_0} + \overline{a_1}\alpha + ... + \overline{a_{m-1}}\alpha^{m-1}$$

with $a_i \in \mathbb{Z}_{p^s}$ $(i = 0, 1, \cdots, m-1)$ and where $\overline{P_m(\xi)}$ is a monic irreducible polynomial of degree m over the Galois field $\mathbb{F}_p$ ($\alpha \in \mathbb{GF}(p^m)$ is a root of $\overline{P_m(\xi)}$).

Three particular cases should be mentioned.

– For $m = 1$, the Galois ring

$$\mathbb{GR}(p^s, 1) = \mathbb{Z}_{p^s}[\xi]/\langle P_1(\xi)\rangle = \mathbb{Z}_{p^s}$$

is nothing but the ring $\mathbb{Z}_{p^s}$ of integers modulo p^s, a ring of characteristic p^s with p^s elements.

– In the case $s = 1$, the Galois ring

$$\mathbb{GR}(p, m) = \mathbb{Z}_p[\xi]/\langle P_m(\xi)\rangle = \mathbb{F}_p[\xi]/\langle P_m(\xi)\rangle = \mathbb{GF}(p^m)$$

is in fact a field, viz. the Galois field $\mathbb{GF}(p^m)$, a field of characteristic p with p^m elements. In this regard, a Galois field is a particular Galois ring.

– Finally, for $s = m = 1$, the Galois ring

$$\mathbb{GR}(p, 1) = \mathbb{Z}_p[\xi]/\langle P_1(\xi)\rangle$$

is a field, viz. the prime field $\mathbb{F}_p$, a field of characteristic p with p elements.

3.2.5. *Isomorphism of two Galois rings*

PROPOSITION 3.4.– Any Galois ring of characteristic p^s and cardinal $(p^s)^m$, with s and m positive integers and p prime number, is isomorphic to an extension $\mathbb{Z}_{p^s}[\xi]/\langle P_m(\xi)\rangle$ of a Galois ring $\mathbb{Z}_{p^s}$, where $P_m(\xi)$ is a monic basic irreducible polynomial of degree m in $\mathbb{Z}_{p^s}[\xi]$. Therefore, two Galois rings $\mathbb{Z}_{p^s}[\xi]/\langle P_m(\xi)\rangle$ and $\mathbb{Z}_{p^s}[\xi]/\langle Q_m(\xi)\rangle$, corresponding to two different monic basic irreducible polynomials $P_m(\xi)$ and $Q_m(\xi)$ of the same degree in $\mathbb{Z}_{p^s}[\xi]$, are isomorphic.

The result, according to which two Galois rings of the same characteristic and the same cardinal are isomorphic, justifies the notation $\mathbb{GR}(p^s, m)$ for denoting any Galois ring of characteristic p^s and cardinal $(p^s)^m$.

3.2.6. *Sub-ring of a Galois ring*

3.2.6.1. *Sub-ring of* $\mathbb{GR}(p^s, m)$

PROPOSITION 3.5.– Any sub-ring of a Galois ring is a Galois ring. The Galois ring $\mathbb{GR}(p^s, \ell)$ is a sub-ring of the Galois ring $\mathbb{GR}(p^s, m)$ if and only if ℓ divides m.

We use the notation $\mathbb{GR}(p^s, \ell) \subset \mathbb{GR}(p^s, m)$ to indicate that $\mathbb{GR}(p^s, \ell)$ is a sub-ring of $\mathbb{GR}(p^s, m)$. The Galois rings $\mathbb{GR}(p^s, \ell)$ and $\mathbb{GR}(p^s, m)$ have the same characteristic p^s. It is important to note that the number of sub-rings of $\mathbb{GR}(p^s, m)$ is equal to the number of positive divisors of m.

3.2.6.2. *Example*

The Galois ring $\mathbb{GR}(p^s, m)$ contains the Galois ring $\mathbb{Z}_{p^s}$ as a sub-ring. Also note that

$$\mathbb{Z}_{p^s} = \mathbb{GR}(p^s, 1!) \subset \mathbb{GR}(p^s, 2!) \subset \mathbb{GR}(p^s, 3!) \subset \cdots$$

since $n!$ divides $(n+1)!$ for $n \in \mathbb{N}_1$. For $s = 1$, we have that

$$\mathbb{F}_p = \mathbb{GF}(p^{1!}) \subset \mathbb{GF}(p^{2!}) \subset \mathbb{GF}(p^{3!}) \subset \cdots$$

in terms of Galois fields.

3.2.7. *adic (p-adic) decomposition*

3.2.7.1. *Two decompositions of an element of* $\mathbb{GR}(p^s, m)$

According to Proposition 3.3, in the Galois ring $\mathbb{GR}(p^s, m) = \mathbb{Z}_{p^s}[\xi]/\langle P_m(\xi)\rangle$, there exists a non-zero element α of order $p^m - 1$ (i.e.

$\alpha^{p^m-1} = 1$) that is a root of the monic basic primitive polynomial $P_m(\xi)$ of degree m over $\mathbb{Z}_{p^s}$ (with $P_m(\xi)$ dividing $\xi^{p^m-1} - 1$ in $\mathbb{Z}_{p^s}[\xi]$) such that

$$\forall a \in \mathbb{GR}(p^s, m) : a = a_0 + a_1\alpha + \cdots + a_{m-1}\alpha^{m-1}$$

$$a_0, a_1, \cdots, a_{m-1} \in \mathbb{Z}_{p^s} \qquad [3.1]$$

where the coefficients a_i $(0 \leq i \leq m-1)$ are unique. Another decomposition of a is given by the following result.

PROPOSITION 3.6.– Let

$$T_m = \{0, \alpha, \cdots, \alpha^{p^m-2}, \alpha^{p^m-1}\} = \{0, 1, \alpha, \cdots, \alpha^{p^m-2}\}$$

be a subset (of cardinal p^m) of $\mathbb{GR}(p^s, m)$. This set is called the *Teichmüller set* of $\mathbb{GR}(p^s, m)$. As an alternative to equation [3.1], we have

$$\forall a \in \mathbb{GR}(p^s, m) : a = t_0 + t_1 p + \cdots + t_{s-1}p^{s-1}$$

$$t_0, t_1, \cdots, t_{s-1} \in T_m \qquad [3.2]$$

where the coefficients t_i $(0 \leq i \leq s-1)$ are unique. In addition, a is a unit if and only if $t_0 \neq 0$ or a zero divisor (including 0) if and only if $t_0 = 0$.

The representation [3.2] of the element a is called *p-adic decomposition* of a. The p-adic representation of an element of the Galois ring $\mathbb{GR}(p^s, m)$ parallels the power representation of an element of the Galois field $\mathbb{GF}(p^m)$ (see the case $s = 1$ below).

3.2.7.2. *Particular cases:* $s = 1$ *and* $s = 2$

– The particular case $s = 1$ corresponds to

$$\mathbb{GR}(p^1, m) = \mathbb{Z}_{p^1}[\xi]/\langle P_m(\xi)\rangle = \mathbb{F}_p[\xi]/\langle P_m(\xi)\rangle = \mathbb{GF}(p^m)$$

and

$$a = t_0$$

where t_0 is an element of the Teichmüller set $\{0, \alpha, \cdots, \alpha^{p^m-1}\}$. In this case, the elements of T_m coincide with the elements of $\mathbb{GF}(p^m)$ expressed in the power representation.

– The case $s = 2$ deserves special attention. In this case, any element a of $\mathbb{GR}(p^2, m)$ can be decomposed as

$$a = t_0 + t_1 p, \quad t_0, t_1 \in T_m$$

which gives

$$a = t_0 + 2t_1, \quad t_0, t_1 \in T_m$$

for $p = 2$.

3.3. Examples and counter-examples of Galois rings

3.3.1. *Counter-examples*

3.3.1.1. *The ring* $\mathbb{Z}_{2^1}[\xi]/\langle \xi^2 \rangle$

The ring $\mathbb{Z}_2[\xi]/\langle \xi^2 \rangle$ is of characteristic 2. Its elements are of the form $a_0 + a_1\alpha$ where a_0 and a_1 belong to $\mathbb{Z}_2$. Therefore, they are $0, 1, \alpha, 1 + \alpha$. The addition and multiplication tables of $\mathbb{Z}_2[\xi]/\langle \xi^2 \rangle$ are given by Tables 3.2 and 3.3, respectively. As already mentioned in 1.1.5.4, the ring $\mathbb{Z}_2[\xi]/\langle \xi^2 \rangle$ is not isomorphic to the ring $\mathbb{Z}_4$ (the two rings have the same multiplication table but have different addition tables).

$+$	0	1	α	$1+\alpha$
0	0	1	α	$1+\alpha$
1	1	0	$1+\alpha$	α
α	α	$1+\alpha$	0	1
$1+\alpha$	$1+\alpha$	α	1	0

Table 3.2. *Addition table for* $\mathbb{Z}_2[\xi]/\langle \xi^2 \rangle$

$\times$	0	1	α	$1+\alpha$
0	0	0	0	0
1	0	1	α	$1+\alpha$
α	0	α	0	α
$1+\alpha$	0	$1+\alpha$	α	1

Table 3.3. *Multiplication table for* $\mathbb{Z}_2[\xi]/\langle \xi^2 \rangle$ *(*α *is such that* $\alpha^2 = 0$*)*

The set $\{0, \alpha\}$ of the two zero divisors of $\mathbb{Z}_2[\xi]/\langle\xi^2\rangle$ forms a principal ideal, but this ideal is not of type $\langle p\rangle$ with p prime. Consequently, the ring $\mathbb{Z}_2[\xi]/\langle\xi^2\rangle$ is not a Galois ring.

Note that the ring $\mathbb{Z}_2[\xi]/\langle\xi^2\rangle$ is isomorphic to the ring $\mathbb{Z}_2[\xi]/\langle 1 + \xi^2\rangle$. Table 3.4 shows the correspondence between the elements of the two rings.

$\mathbb{Z}_2[\xi]/\langle\xi^2\rangle$	0	1	α	$1+\alpha$
$\mathbb{Z}_2[\xi]/\langle 1+\xi^2\rangle$	0	1	$1+\beta$	β

Table 3.4. *Correspondence between the elements* $(0, 1, \alpha, 1 + \alpha)$ *and* $(0, 1, \beta, 1 + \beta)$ *of* $\mathbb{Z}_2[\xi]/\langle\xi^2\rangle$ *and* $\mathbb{Z}_2[\xi]/\langle 1 + \xi^2\rangle$*, respectively*

3.3.1.2. *The ring* $\mathbb{Z}_{2^1}[\xi]/\langle\xi + \xi^2\rangle$

As for $\mathbb{Z}_2[\xi]/\langle\xi^2\rangle$, the characteristic of the ring $\mathbb{Z}_2[\xi]/\langle\xi + \xi^2\rangle$ is equal to 2. Its elements are $0, 1, \alpha, 1 + \alpha$, and the corresponding addition and multiplication tables are given by Tables 3.5 and 3.6, respectively. The ring $\mathbb{Z}_2[\xi]/\langle\xi + \xi^2\rangle$ has one unit (1), three zero divisors (0, α and $1 + \alpha$) and two non-trivial ideals ($\{0, \alpha\}$ and $\{0, 1 + \alpha\}$). The three zero divisors do not constitute an ideal so that $\mathbb{Z}_2[\xi]/\langle\xi + \xi^2\rangle$ is not a Galois ring.

$+$	0	1	α	$1+\alpha$
0	0	1	α	$1+\alpha$
1	1	0	$1+\alpha$	α
α	α	$1+\alpha$	0	1
$1+\alpha$	$1+\alpha$	α	1	0

Table 3.5. *Addition table for* $\mathbb{Z}_2[\xi]/\langle\xi + \xi^2\rangle$

$\times$	0	1	α	$1+\alpha$
0	0	0	0	0
1	0	1	α	$1+\alpha$
α	0	α	α	0
$1+\alpha$	0	$1+\alpha$	0	$1+\alpha$

Table 3.6. *Multiplication table for* $\mathbb{Z}_2[\xi]/\langle\xi + \xi^2\rangle$ *(*α *is such that* $\alpha + \alpha^2 = 0$*)*

Observe that $\mathbb{Z}_2[\xi]/\langle\xi + \xi^2\rangle$ is neither isomorphic to $\mathbb{Z}_2[\xi]/\langle\xi^2\rangle$ nor to $\mathbb{Z}_4$: both the addition and multiplication tables of $\mathbb{Z}_2[\xi]/\langle\xi + \xi^2\rangle$ differ from those of the ring $\mathbb{Z}_4$; furthermore, the rings $\mathbb{Z}_2[\xi]/\langle\xi^2\rangle$ and $\mathbb{Z}_2[\xi]/\langle\xi + \xi^2\rangle$ have the same addition table but different multiplication tables. The ring $\mathbb{Z}_2[\xi]/\langle\xi+\xi^2\rangle$ is isomorphic to the direct product $\mathbb{F}_2 \times \mathbb{F}_2$ via the correspondence

$$0 \leftrightarrow (0,0), \quad 1 \leftrightarrow (1,1), \quad \alpha \leftrightarrow (0,1), \quad 1+\alpha \leftrightarrow (1,0)$$

(for the element (a,b) of $\mathbb{F}_2 \times \mathbb{F}_2$, a belongs to the first field $\mathbb{F}_2$ and b to the second one) and

$$(a,b) + (a',b') = (a+a', b+b'), \quad (a,b) \times (a',b') = (aa', bb')$$

(where the additions of type $a + a'$ and the multiplications of type aa' are in $\mathbb{F}_2$). More generally, one can prove that the ring $\mathbb{Z}_p[\xi]/\langle\xi + \xi^2\rangle$ is isomorphic to the direct product $\mathbb{F}_p \times \mathbb{F}_p$.

3.3.1.3. *The ring $\mathbb{Z}_{2^1}[\xi]/\langle\xi + \xi^3\rangle$*

The ring $\mathbb{Z}_2[\xi]/\langle\xi+\xi^3\rangle$ is of characteristic $2^1 = 2$. Its $(2^1)^3 = 8$ elements, of the form $a_0 + a_1\alpha + a_2\alpha^2$ with $a_0, a_1, a_2 \in \mathbb{Z}_2$, are

$$0,\ 1,\ a = \alpha,\ b = \alpha^2,\ c = 1+\alpha,\ d = 1+\alpha^2,\ e = \alpha+\alpha^2$$

$$f = 1+\alpha+\alpha^2$$

×	0	1	a	b	c	d	e	f
0	0	1	a	b	c	d	e	f
1	1	0	c	d	a	b	f	e
a	a	c	0	e	1	f	b	d
b	b	d	e	0	f	1	a	c
c	c	a	1	f	0	e	d	b
d	d	b	f	1	e	0	c	a
e	e	f	b	a	d	c	0	1
f	f	e	d	c	b	a	1	0

Table 3.7. *Addition table for the rings $\mathbb{Z}_2[\xi]/\langle\xi+\xi^3\rangle$ and $\mathbb{Z}_2[\xi]/\langle 1+\xi^3\rangle$ whose elements are $0,\ 1,\ a = \alpha,\ b = \alpha^2,\ c = 1+\alpha,\ d = 1+\alpha^2$, $e = \alpha+\alpha^2$ and $f = 1+\alpha+\alpha^2$*

The corresponding addition and multiplication tables are given by Tables 3.7 and 3.8. Table 3.8 shows that $\mathbb{Z}_2[\xi]/\langle\xi+\xi^3\rangle$ has two units (1 and $1+\alpha+\alpha^2$), six zero divisors (0, α, α^2, $1+\alpha$, $1+\alpha^2$ and $\alpha+\alpha^2$) and four non-trivial ideals

($\{0, 1+\alpha^2\}$, $\{0, \alpha+\alpha^2\}$, $\{0, \alpha, \alpha^2, \alpha+\alpha^2\}$ and $\{0, 1+\alpha, 1+\alpha^2, \alpha+\alpha^2\}$). These four ideals are principal ideals. The ring $\mathbb{Z}_2[\xi]/\langle\xi + \xi^3\rangle$ is not a Galois ring since the six zero divisors do not form an ideal. Finally, note that the ring $\mathbb{Z}_2[\xi]/\langle\xi + \xi^3\rangle$ is isomorphic to the direct product $\mathbb{F}_2 \times \mathbb{Z}_4$, the elements of $\mathbb{F}_2 \times \mathbb{Z}_4$ being (a, b) with $a \in \mathbb{F}_2$ and $b \in \mathbb{Z}_4$.

×	0	1	a	b	c	d	e	f
0	0	0	0	0	0	0	0	0
1	0	1	a	b	c	d	e	f
a	0	a	b	a	e	0	e	b
b	0	b	a	b	e	0	e	a
c	0	c	e	e	d	d	0	c
d	0	d	0	0	d	d	0	d
e	0	e	e	e	0	0	0	e
f	0	f	b	a	c	d	e	1

Table 3.8. *Multiplication table for the ring $\mathbb{Z}_2[\xi]/\langle\xi + \xi^3\rangle$ whose elements are 0, 1, $a = \alpha$, $b = \alpha^2$, $c = 1 + \alpha$, $d = 1 + \alpha^2$, $e = \alpha + \alpha^2$ and $f = 1 + \alpha + \alpha^2$ (α is such that $\alpha + \alpha^3 = 0$)*

×	0	1	a	b	c	d	e	f
0	0	0	0	0	0	0	0	0
1	0	1	a	b	c	d	e	f
a	0	a	b	1	e	c	d	f
b	0	b	1	a	d	e	c	f
c	0	c	e	d	d	e	c	0
d	0	d	c	e	e	c	d	0
e	0	e	d	c	c	d	e	0
f	0	f	f	f	0	0	0	f

Table 3.9. *Multiplication table for the ring $\mathbb{Z}_2[\xi]/\langle 1 + \xi^3\rangle$ whose elements are 0, 1, $a = \alpha$, $b = \alpha^2$, $c = 1 + \alpha$, $d = 1 + \alpha^2$, $e = \alpha + \alpha^2$ and $f = 1 + \alpha + \alpha^2$ (α is such that $1 + \alpha^3 = 0$)*

3.3.1.4. *The ring $\mathbb{Z}_{2^1}[\xi]/\langle 1 + \xi^3\rangle$*

The ring $\mathbb{Z}_2[\xi]/\langle 1 + \xi^3\rangle$ is a ring of characteristic 2 and cardinal 8. The addition and multiplication tables for the eight elements

$$0,\ 1,\ a = \alpha,\ b = \alpha^2,\ c = 1 + \alpha,\ d = 1 + \alpha^2,\ e = \alpha + \alpha^2$$

$$f = 1 + \alpha + \alpha^2$$

of $\mathbb{Z}_2[\xi]/\langle 1+\xi^3\rangle$ are given by Tables 3.7 and 3.9, respectively. The ring $\mathbb{Z}_2[\xi]/\langle 1+\xi^3\rangle$ is not isomorphic to $\mathbb{Z}_2[\xi]/\langle \xi+\xi^3\rangle$ (the two rings have the same addition table, but their multiplication tables differ). Table 3.9 shows that $\mathbb{Z}_2[\xi]/\langle 1+\xi^3\rangle$ has five zero divisors (0, $1+\alpha$, $1+\alpha^2$, $\alpha+\alpha^2$ and $1+\alpha+\alpha^2$). These five zero divisors do not form an ideal of $\mathbb{Z}_2[\xi]/\langle 1+\xi^3\rangle$. Therefore, $\mathbb{Z}_2[\xi]/\langle 1+\xi^3\rangle$ is not a Galois ring.

3.3.1.5. *Remark*

It is clear that the polynomials $P_m(\xi)$ occurring in the four preceding counter-examples are not monic basic irreducible polynomials in $\mathbb{Z}_2[\xi]$. This confirms that the rings 3.3.1.1 to 3.3.1.4 are not Galois rings.

3.3.2. *Examples*

3.3.2.1. *The Galois ring* $\mathbb{GR}(2^2, 1) = \mathbb{Z}_4$

In the case where $m = 1$, the ring $\mathbb{GR}(p^s, 1)$ is isomorphic to the base ring $\mathbb{Z}_{p^s}$. Let us consider the case $p = s = 2$ that corresponds to the ring $\mathbb{Z}_4$, a ring of characteristic 4 with four elements. It is easy to check that every element a of $\mathbb{Z}_4$ can be written as

$$a = t_0 + 2 \times t_1, \quad t_0, t_1 \in \mathbb{Z}_2$$

where the operations $+$ and $\times$ are performed in $\mathbb{Z}_4$ (here $T_1 = \mathbb{Z}_2$). The zero divisors (0 and 2) of the unitary ring $\mathbb{Z}_4$ form a principal ideal $\langle 2\rangle$. Therefore, $\mathbb{Z}_4$ is a Galois ring.

3.3.2.2. *The Galois ring* $\mathbb{GR}(2^1, 4) = \mathbb{Z}_{2^1}[\xi]/\langle 1+\xi+\xi^4\rangle$

The Galois field $\mathbb{GF}(2^4) = \mathbb{F}_2[\xi]/\langle 1+\xi+\xi^4\rangle$ of characteristic 2 and cardinal 16, see 2.3.8.6, is also a Galois ring in the sense that $1+\xi+\xi^4$ is obviously a monic basic primitive polynomial over $\mathbb{Z}_2$ and the unique zero divisor 0 constitutes a principal ideal (here $\langle 0\rangle$ plays the role of $\langle p = 2\rangle$). Therefore, $\mathbb{GF}(2^4) = \mathbb{GR}(2^1, 4)$.

3.3.2.3. *The Galois ring* $\mathbb{GR}(2^2, 2) = \mathbb{Z}_{2^2}[\xi]/\langle 1+\xi+\xi^2\rangle$

The polynomial $1+\xi+\xi^2$ is clearly a monic basic primitive polynomial over $\mathbb{Z}_4$. The ring $\mathbb{GR}(2^2, 2) = \mathbb{Z}_4[\xi]/\langle 1+\xi+\xi^2\rangle$ (not to be confused with the field $\mathbb{GF}(2^2) = \mathbb{F}_2[\xi]/\langle 1+\xi+\xi^2\rangle)$ is of characteristic $2^2 = 4$ and has $(2^2)^2 = 16$ elements. The elements

$$a = a_0 + a_1\alpha, \quad a_0, a_1 \in \mathbb{Z}_4$$

of $\mathbb{Z}_4[\xi]/\langle 1+\xi+\xi^2\rangle$ are

$$0,\ 3+3\alpha \equiv \alpha^2,\ 2+2\alpha \equiv 2\alpha^2,\ \alpha$$

$$3 \equiv \alpha+\alpha^2,\ 2\alpha,\ 3+\alpha \equiv 2\alpha+\alpha^2,\ 2 \equiv 2\alpha+2\alpha^2$$

$$3\alpha,\ 3+2\alpha \equiv 3\alpha+\alpha^2,\ 2+\alpha \equiv 3\alpha+2\alpha^2,\ 1 \equiv 3\alpha+3\alpha^2 = \alpha^3$$

$$1+\alpha \equiv 3\alpha^2,\ 1+2\alpha \equiv \alpha+3\alpha^2,\ 2+3\alpha \equiv \alpha+2\alpha^2$$

$$1+3\alpha \equiv 2\alpha+3\alpha^2$$

modulo 4 and modulo $1+\alpha+\alpha^2 = 0$ (the element α, root of $1+\xi+\xi^2$, is of order 2^2-1). The set $\{0, 2, 2\alpha, 2+2\alpha\}$ of all zero divisors constitutes a principal ideal $\langle 2\rangle$ of $\mathbb{Z}_4[\xi]/\langle 1+\xi+\xi^2\rangle$. Therefore, $\mathbb{Z}_4[\xi]/\langle 1+\xi+\xi^2\rangle$ is a Galois ring.

In the p-adic representation, every element a of the ring $\mathbb{Z}_4[\xi]/\langle 1+\xi+\xi^2\rangle$ can be written as

$$a = t_0 + 2 \times t_1, \quad t_0, t_1 \in T_2$$

(with addition $+$ and multiplication $\times$ in $\mathbb{Z}_4$) where T_2 is the Teichmüller set

$$T_2 = \{0, \alpha, \alpha^2, 1\} = \{0, \alpha, \alpha^2, \alpha^3\} = \{0, \alpha, 3+3\alpha, 1\}$$

The 16 elements $t_0 + 2\times t_1$ of $\mathbb{Z}_4[\xi]/\langle 1+\xi+\xi^2\rangle$ are reported in Table 3.10 in terms of t_0 and t_1.

3.3.2.4. *The Galois ring* $\mathbb{GR}(2^2, 3) = \mathbb{Z}_{2^2}[\xi]/\langle 3+\xi+2\xi^2+\xi^3\rangle$

The monic polynomial $P_3(\xi) = 3+\xi+2\xi^2+\xi^3$ in $\mathbb{Z}_{2^2}[\xi]$ admits the image $\overline{P_3(\xi)} = 1+\xi+\xi^3$ in $\mathbb{Z}_2[\xi]$, an irreducible polynomial over $\mathbb{Z}_2$. Therefore, $P_3(\xi)$ is a monic basic irreducible polynomial. Thus, the ring $\mathbb{GR}(2^2, 3)$ is a Galois ring of characteristic $2^2 = 4$ with $(2^2)^3 = 64$ elements. Let α be a root of $P_3(\xi)$. This root is an element of order $2^3 - 1 = 7$ of $\mathbb{GR}(2^2, 3)$. Indeed, the various powers of α are

$$\alpha, \quad \alpha^2, \quad \alpha^3 = 1 + 3\alpha + 2\alpha^2$$

$$\alpha^4 = 2 + 3\alpha + 3\alpha^2, \quad \alpha^5 = 3 + 3\alpha + \alpha^2$$

$$\alpha^6 = 1 + 2\alpha + \alpha^2, \quad \alpha^7 = 1$$

$t_1 \rightarrow$ $t_0 \downarrow$	0	1	α	$3+3\alpha$
0	0	2	2α	$2+2\alpha$
1	1	3	$1+2\alpha$	$3+2\alpha$
α	α	$2+\alpha$	3α	$2+3\alpha$
$3+3\alpha$	$3+3\alpha$	$1+3\alpha$	$3+\alpha$	$1+\alpha$

Table 3.10. *Elements of $\mathbb{GR}(2^2, 2) = \mathbb{Z}_{2^2}[\xi]/\langle 1 + \xi + \xi^2 \rangle$: the element $t_0 + 2 \times t_1$ stands at the intersection of the row t_0 and the column t_1 (t_0 and t_1 belong to the Teichmüller set T_2)*

The Teichmüller set T_3 of the Galois ring $\mathbb{GR}(2^2, 3)$ is

$$T_3 = \{0, 1, \alpha, \alpha^2, \alpha^3, \alpha^4, \alpha^5, \alpha^6\} = \{0, \alpha, \alpha^2, \alpha^3, \alpha^4, \alpha^5, \alpha^6, \alpha^7\}$$

and every element a of $\mathbb{GR}(2^2, 3)$ can be written as

$$a = t_0 + 2t_1, \quad t_0, t_1 \in T_3$$

in a unique way (various writings of the elements of the Teichmüller set T_3 are given in Table 3.11).

3.3.2.5. *The Galois ring $\mathbb{GR}(2^2, m)$*

The Galois ring $\mathbb{GR}(2^2, m)$ is of interest in quantum information for describing a system of m qubits ($\mathbb{GR}(2^2, m)$ is often denoted as R_{4^m} in quantum information). It is of characteristic $2^2 = 4$, has $(2^2)^m = 4^m$ elements and corresponds to

$$\mathbb{GR}(2^2, m) = \mathbb{Z}_{2^2}[\xi]/\langle P_m(\xi) \rangle$$

where $P_m(\xi)$ is a monic basic irreducible polynomial of degree m in $\mathbb{Z}_{2^2}[\xi]$ (its image under the homomorphism $\mathbb{Z}_{2^2}[\xi] \to \mathbb{Z}_2[\xi]$ is an irreducible polynomial over $\mathbb{Z}_2$). The Galois ring $\mathbb{GR}(2^2, m)$ is an extension of degree m of the ring $\mathbb{Z}_{2^2}$.

Power Form	Polynomial Form	$[a_0a_1a_2]_4$ Form	$[\overline{a}_0\overline{a}_1\overline{a}_2]_2$ Form
0	0	$[0\,0\,0]_4$	$[0\,0\,0]_2$
α^7	1	$[1\,0\,0]_4$	$[1\,0\,0]_2$
α^1	α^1	$[0\,1\,0]_4$	$[0\,1\,0]_2$
α^2	α^2	$[0\,0\,1]_4$	$[0\,0\,1]_2$
α^3	$1+3\alpha+2\alpha^2$	$[1\,3\,2]_4$	$[1\,1\,0]_2$
α^4	$2+3\alpha+3\alpha^2$	$[2\,3\,3]_4$	$[0\,1\,1]_2$
α^5	$3+3\alpha+\alpha^2$	$[3\,3\,1]_4$	$[1\,1\,1]_2$
α^6	$1+2\alpha+\alpha^2$	$[1\,2\,1]_4$	$[1\,0\,1]_2$

Table 3.11. *Elements of the Teichmüller set T_3 for the ring $\mathbb{GR}(2^2,3) = \mathbb{Z}_{2^2}[\xi]/\langle 3+\xi+2\xi^2+\xi^3\rangle$: the element $a_0+a_1\alpha+a_2\alpha^2$ of the ring $\mathbb{GR}(2^2,3)$ is described by $[a_0a_1a_2]_4$ with $a_0,a_1,a_2 \in \mathbb{Z}_{2^2}$ whereas $[\overline{a}_0\overline{a}_1\overline{a}_2]_2$ with $\overline{a}_0,\overline{a}_1,\overline{a}_2 \in \mathbb{Z}_2$ gives the corresponding element in the field $\mathbb{GF}(2^3)$ obtained via the homomorphism $\mathbb{GR}(2^2,3) = \mathbb{Z}_{2^2}[\xi]/\langle 3+\xi+2\xi^2+\xi^3\rangle \to \mathbb{GF}(2^3) = \mathbb{F}_2[\xi]/\langle 1+\xi+\xi^3\rangle$*

There exists a non-zero element α of order 2^m-1 ($\alpha^{2^m-1}=1$) in $\mathbb{GR}(2^2,m)$. The element α (an n-th root of unity with $n=2^m-1$) is a root of $P_m(\xi)$. Any element a of $\mathbb{GR}(2^2,m)$ can be uniquely expressed in the polynomial form

$$a = a_0 + a_1\alpha + \cdots + a_{m-1}\alpha^{m-1}, \quad a_0, a_1, \cdots, a_{m-1} \in \mathbb{Z}_{2^2}$$

or in the 2-adic form

$$a = t_0 + 2t_1, \quad t_0, t_1 \in T_m$$

where

$$T_m = \{0, \alpha, \cdots, \alpha^{2^m-2}, \alpha^{2^m-1}\} = \{0, 1, \alpha, \cdots, \alpha^{2^m-2}\}$$

is the Teichmüller set of the Galois ring $\mathbb{GR}(2^2,m)$. Note that $a^{2m}=t_0$ and the elements $a=t_0+2t_1$ with $t_0 \neq 0$ are units.

As a particular case, we have $\mathbb{GR}(2^2,1) = \mathbb{Z}_4$ for $m=1$. For $\mathbb{GR}(2^2,m) = \mathbb{Z}_{2^2}[\xi]/\langle P_m(\xi)\rangle$ with $m \geq 2$, the following cases are of importance in quantum information:

– the case $m=2$ (for two qubits) with $P_2(\xi) = 1+\xi+\xi^2$;

– the case $m=3$ (for three qubits) with $P_3(\xi) = 3+\xi+2\xi^2+\xi^3$;

– the case $m=4$ (for four qubits) with $P_4(\xi) = 1+3\xi+2\xi^2+\xi^4$.

3.4. The application trace for a Galois ring

3.4.1. *Generalized Frobenius automorphism and trace*

DEFINITION 3.4.– The map

$$\phi : \mathbb{GR}(p^s, m) \to \mathbb{GR}(p^s, m)$$
$$a \mapsto \phi(a)$$

with

$$a = a_0 + a_1 x + \ldots + a_{m-1}\, \alpha^{m-1}$$
$$\phi(a) = a_0 + a_1 x^p + \ldots + a_{m-1}\, \alpha^{(m-1)p}$$

defines an automorphism of $\mathbb{GR}(p^s, m)$ called generalized Frobenius automorphism. The *trace* of $a \in \mathbb{GR}(p^s, m)$ is defined by the surjective map

$$\mathrm{Tr} : \mathbb{GR}(p^s, m) \to \mathbb{Z}_{p^s}$$
$$a \mapsto \mathrm{Tr}(a) = a + \phi(a) + \phi^2(a) + \cdots + \phi^{m-1}(a)$$

in terms of the generalized Frobenius map ϕ.

By introducing

$$\phi^0(a) = a,\ \phi^1(a) = \phi(a),\ \phi^2(a) = \phi(\phi^1(a)),\ \cdots,\ \phi^{m-1}(a)$$
$$= \phi(\phi^{m-2}(a))$$

then $\mathrm{Tr}(a)$ can be rewritten as

$$\mathrm{Tr}(a) = \sum_{k=0}^{m-1} \phi^k(a)$$

Note that the map

$$\phi : \mathbb{GR}(p^s, m) \to \mathbb{GR}(p^s, m)$$
$$a \mapsto \phi(a)$$

generalizes the Frobenius map

$$\begin{aligned}\sigma : \mathbb{GF}(p^m) &\to \mathbb{GF}(p^m)\\ x &\mapsto \sigma(x) = x^p\end{aligned}$$

In fact, for $s = 1$, we again find the definitions of the Frobenius map and of the trace for Galois fields (with $\phi = \sigma$).

The case $m = 1$ corresponds to $\mathbb{GR}(p^s, 1) = \mathbb{Z}_{p^s}$ for which $\phi(a) = a$ and $\mathrm{Tr}(a) = a$ for any a in $\mathbb{Z}_{p^s}$.

As an illustration, for the Galois ring $\mathbb{GR}(2^2, 2) = \mathbb{Z}_{2^2}[\xi]/\langle 1 + \xi + \xi^2\rangle$, we have

$$\begin{aligned}\phi : \mathbb{GR}(2^2, 2) &\to \mathbb{GR}(2^2, 2)\\ t_0 + 2t_1 &\mapsto \phi(t_0 + 2t_1) = {t_0}^2 + 2{t_1}^2\end{aligned}$$

and

$$\begin{aligned}\mathrm{Tr} : \mathbb{GR}(2^2, 2) &\to \mathbb{Z}_{2^2}\\ t_0 + 2t_1 &\mapsto \mathrm{Tr}(t_0 + 2t_1) = t_0 + 2t_1 + {t_0}^2 + 2{t_1}^2\end{aligned}$$

where t_0 and t_1 belong to the Teichmüller set T_2 of the ring $\mathbb{GR}(2^2, 2)$.

3.4.2. *Elementary properties of the trace*

PROPOSITION 3.7.– The following properties

– Property 1:

$$\forall a \in \mathbb{GR}(p^s, m) : \mathrm{Tr}(a) \in \mathbb{Z}_{p^s}$$

– Property 2:

$$\forall a \in \mathbb{GR}(p^s, m),\ \forall b \in \mathbb{GR}(p^s, m) : \mathrm{Tr}(a + b) = \mathrm{Tr}(a) + \mathrm{Tr}(b)$$

– Property 3:

$$\forall k \in \mathbb{Z}_{p^s},\ \forall a \in \mathbb{GR}(p^s, m) : \mathrm{Tr}(ka) = k\mathrm{Tr}(a)$$

– Property 4:

$$\forall a \in \mathbb{GR}(p^s, m) : \mathrm{Tr}(\phi(a)) = \mathrm{Tr}(a)$$

are reminiscent of some properties of the field $\mathbb{GF}(p^m) = \mathbb{GR}(p^1, m)$.

3.5. Characters of a Galois ring

For a Galois ring $\mathbb{GR}(p^s, m)$, there are two group structures corresponding to the addition law $+$ and the multiplication law $\times$ of the ring:

– the additive group $(\mathbb{GR}(p^s, m), +)$ spanned by all the elements of the ring $\mathbb{GR}(p^s, m)$;

– the multiplicative group $(\mathbb{GR}(p^s, m)^*, \times)$ spanned by all the units of the ring $\mathbb{GR}(p^s, m)$.

Here $\mathbb{GR}(p^s, m)^*$ stands for $\mathbb{GR}(p^s, m) \setminus S$, where S is the set of zero divisors, including the trivial zero divisor, of $\mathbb{GR}(p^s, m)$. These two group structures give rise to two kinds of characters: the *additive characters* $\chi_b(a)$ and the *multiplicative characters* $\psi_k(a)$. They are defined in a way similar to those for a Galois field $\mathbb{GF}(p^m)$. For instance, the additive character vector χ_b is defined by

$$\chi_b : \mathbb{GR}(p^s, m) \to \mathbb{C}^*$$

$$a \mapsto \chi_b(a) = \mathrm{e}^{\mathrm{i}\frac{2\pi}{p^s}\mathrm{Tr}(ab)}, \quad b \in \mathbb{GR}(p^s, m)$$

Observe that for $s = 1$, we recover the definition of additive characters given for a Galois field $\mathbb{GF}(p^m)$. A similar definition can be given for the multiplicative character vector ψ_k of $\mathbb{GR}(p^s, m)$.

As an illustration, for the Galois ring $\mathbb{Z}_{2^2} = \mathbb{GR}(2^2, 1)$, we have

$$\chi_b(a) = \mathrm{i}^{ab}, \quad a, b \in \mathbb{Z}_{2^2}$$

for the additive character vectors χ_b and

$$\psi_k(a) = \mathrm{e}^{\mathrm{i}\pi ak}, \quad a, k \in \mathbb{Z}_2$$

for the multiplicative character vectors ψ_k.

3.6. Gaussian sums over Galois rings

3.6.1. *Gauss sum over* $\mathbb{GR}(p^s, m)$

As for a Galois field, we can define a Gaussian sum for a Galois ring. The Gaussian sum $G_m(\psi_k, \chi_b)$ for the Galois ring $\mathbb{GR}(p^s, m)$ is defined by

$$G_m(\psi_k, \chi_b) = \sum_{a \in \mathbb{GR}(p^s, m)^*} \psi_k(a)\chi_b(a)$$

where ψ_k is an arbitrary multiplicative character vector of $(\mathbb{GR}(p^s, m)^*, \times)$ and χ_b an arbitrary additive character vector of $(\mathbb{GR}(p^s, m), +)$, and the summation on a runs over all the units of $\mathbb{GR}(p^s, m)$. The evaluation of the sum $G_m(\psi_k, \chi_b)$ for a Galois ring is more involved than the one for a Galois field. As an example, we have

$$G_m(\psi_k, \chi_0) = 0 \text{ for } \psi_k \neq \psi_0$$

where ψ_0 is the trivial multiplicative character vector of $(\mathbb{GR}(p^s, m)^*, \times)$ and χ_0 the trivial additive character vector of $(\mathbb{GR}(p^s, m), +)$.

3.6.2. *Weil sum over* $\mathbb{GR}(p^s, m)$

Let χ_a be an additive character of the Galois ring $\mathbb{GR}(p^s, m)$ and Γ be the map

$$\Gamma : \mathbb{GR}(p^s, m) \to \mathbb{C}$$
$$a \mapsto \Gamma(a) = \sum_{t \in T_m} \chi_a(t)$$

where T_m is the Teichmüller set of $\mathbb{GR}(p^s, m)$. The sum

$$\Gamma(a) = \sum_{t \in T_m} \chi_a(t) = \sum_{t \in T_m} \mathrm{e}^{\mathrm{i}\frac{2\pi}{p^s}\mathrm{Tr}(at)}, \quad a \in \mathbb{GR}(p^s, m)$$

is called a Weil sum.

As an example, we have

$$\left|\sum_{t\in T_m} \mathrm{e}^{\mathrm{i}\frac{\pi}{2}\mathrm{Tr}(at)}\right| = \begin{cases} 0 \text{ if } a \in 2T_m,\ a \neq 0 \\ 2^m \text{ if } a = 0 \\ \sqrt{2^m} \text{ otherwise} \end{cases}$$

for the Galois ring $\mathbb{GR}(2^2, m)$.

4

Mutually Unbiased Bases

According to Moore's law, the size of electronic and spintronic devices for a classical computer should approach 10 nm in 2018–2020, i.e. the scale where quantum effects are visible, a fact in favor of a quantum computer. This explains the growing interest for a new field, namely the field of quantum information and quantum computation. Such a field, which started in the 1980s, is at the crossroads of quantum mechanics, discrete mathematics and informatics with the aim of building a quantum computer. We note in passing that, even in the case where the aim would not be reached, physics, mathematics, informatics and engineering will greatly benefit from the enormous amount of work along this line.

In a quantum computer, classical bits (0 and 1) are replaced by quantum bits or qubits (that interpolate in some sense between 0 and 1). A qubit is a vector $|\psi\rangle$ in the two-dimensional Hilbert space $\mathbb{C}^2$:

$$|\psi\rangle = x|0\rangle + y|1\rangle, \quad x \in \mathbb{C}, \quad y \in \mathbb{C}, \quad |x|^2 + |y|^2 = 1$$

where $|0\rangle$ and $|1\rangle$ are the elements of an orthonormal basis in this space. The result of a measurement of $|\psi\rangle$ is not deterministic since it gives $|0\rangle$ or $|1\rangle$ with the probability $|x|^2$ or $|y|^2$, respectively. The consideration of N qubits leads to work in the 2^N-dimensional Hilbert space $\mathbb{C}^{2^N}$. Note that the notion of qubit, corresponding to $\mathbb{C}^2$, is a particular case of the one of qudit corresponding to $\mathbb{C}^d$ (d not necessarily in the form 2^N). A system of N qudits is associated with the Hilbert space $\mathbb{C}^{d^N}$. In this connection, the techniques developed for finite-dimensional Hilbert spaces are of paramount importance in quantum computation and quantum calculation.

From a formal point of view, a quantum computer can be considered as a set of qubits, the state of which can be (controlled and) manipulated via unitary transformations. These transformations correspond to the product of elementary unitary operators called quantum gates acting on one or two qubits. Measurement of the qubits coming out from a circuit of quantum gates yields the result of a (quantum) computation. In other words, a realization of quantum information processing can be performed by preparing a quantum system in a quantum state, then submitting this state to a unitary transformation and, finally, reading the outcome from a measurement.

Unitary operator bases of the Hilbert space $\mathbb{C}^d$ are of pivotal importance for quantum information and quantum computation as well as for quantum mechanics in general. The interest for unitary operator bases started with the seminal work by Schwinger. Among such bases, *mutually unbiased bases* (MUBs) play a key role in quantum information and quantum computation (two distinct orthonormal bases B_a and B_b of $\mathbb{C}^d$ are said to be unbiased if and only if the inner product of any vector of one basis by any vector of the other basis has a modulus independent of the two chosen vectors).

There exist numerous ways of constructing sets of MUBs. Most of them are based on the Fourier analysis over Galois fields and Galois rings, discrete Wigner distribution, generalized Pauli spin matrices, Latin squares, finite and projective geometries, convex polytopes, complex projective 2-designs, angular momentum theory and Lie-like methods, and discrete phase states. In this chapter, from quantum theory of angular momentum theory (or, in mathematical terms, from the Lie algebra A_1 of the group $\mathrm{SU}(2)$ or $\mathrm{SL}(2, \mathbb{C})$), we shall derive a formula for a complete set of MUBs in dimension p with p prime. Moreover, we shall construct complete sets of MUBs in dimension p^m with p prime and m positive integer from the additive characters of the field $\mathbb{GF}(p^m)$ for p odd and of the ring $\mathbb{GR}(2^2, m)$ for $p = 2$.

An exhaustive, although incomplete, list of references for this chapter is given in the section *Theoretical physics literature: MUBs* of the bibliography. Each reference in the list is annotated with some key words inside [ ] to characterize the content of the referenced work.

4.1. Generalities

4.1.1. *Unbiased bases*

DEFINITION 4.1.– Let $\mathbb{C}^d$ be the Hilbert space of dimension d over $\mathbb{C}$ and B_a and B_b be two distinct orthonormal bases

$$B_a = \{|a\alpha\rangle \mid \alpha = 0, 1, \cdots, d-1\}$$

$$B_b = \{|b\beta\rangle \mid \beta = 0, 1, \cdots, d-1\}$$

of $\mathbb{C}^d$. The bases B_a and B_b ($a \neq b$) are said to be unbiased if and only if

$$\forall \alpha \in \mathbb{Z}_d, \ \forall \beta \in \mathbb{Z}_d : |\langle a\alpha|b\beta\rangle| = \frac{1}{\sqrt{d}}$$

where $\langle \mid \rangle$ denotes the inner product of $\mathbb{C}^d$. In other words, the inner product $\langle a\alpha|b\beta\rangle$ has a modulus independent of α and β. The relation

$$|\langle a\alpha|b\beta\rangle| = \delta(a, b)\delta(\alpha, \beta) + [1 - \delta(a, b)]\frac{1}{\sqrt{d}}$$

makes it possible to describe both the cases $B_a = B_b$ and $B_a \neq B_b$.

4.1.2. *Example:* $d = 2$

Let H_2 be the Hilbert space (isomorphic to $\mathbb{C}^2$) spanned by two orthonormal vectors $|0\rangle$ and $|1\rangle$. The three bases

$$B_0 : \frac{1}{\sqrt{2}}(|0\rangle + |1\rangle), \quad \frac{1}{\sqrt{2}}(|0\rangle - |1\rangle)$$

$$B_1 : \frac{1}{\sqrt{2}}(|0\rangle + \mathrm{i}|1\rangle), \quad \frac{1}{\sqrt{2}}(|0\rangle - \mathrm{i}|1\rangle)$$

$$B_2 : |0\rangle, \quad |1\rangle$$

constitute a set of three MUBs. Such bases are familiar bases in quantum information for describing qubits.

4.1.3. *Interests of MUBs for quantum mechanics*

MUBs are of relevance to advanced quantum mechanics. A significance of MUBs in terms of quantum measurements can be seen as follows. Let A and B be two non-degenerate (i.e. with multiplicity-free eigenvalues) observables of a quantum system with the Hilbert space $\mathbb{C}^d$ of dimension d. Suppose that the eigenvectors of A and B yield two unbiased bases B_a and B_b, respectively. When the quantum system is prepared in an eigenvector $|b\beta\rangle$ of the observable B, no information can be obtained from a measure of the observable A. This result follows from the development in the basis B_a of any vector of the basis B_b

$$|b\beta\rangle = \sum_{\alpha=0}^{d-1} |a\alpha\rangle\langle a\alpha|b\beta\rangle$$

which shows that the d probabilities

$$|\langle a\alpha|b\beta\rangle|^2 = \frac{1}{d}, \quad \alpha, \beta = 0, 1, \cdots, d-1$$

of obtaining any state vector $|a\alpha\rangle$ in a measure of A are equal.

Two such observables A and B are said to be complementary (Bohr's principle of complementarity introduced in the early days of quantum mechanics): a precise knowledge of one of them implies a total uncertainty of the other (or all possible results of measurements of the other one are equally probable). Such observables are represented by operators that do not commute. The most familiar example is for d infinite. The position $A = x$ and the momentum $B = p_x$ (along the x-direction) are complementary observables. They satisfy the commutation relations

$$[x, p_x] = \mathrm{i}\hbar$$

where $\hbar$ is the Planck constant. From these commutation relations, it is possible to derive the Heisenberg uncertainty inequalities. These inequalities mean that a precise knowledge of x yields a complete indeterminacy of p_x and vice versa.

Sets of MUBs play an important role in the theory of quantum mechanics as for the discrete Wigner function, for the solution of the Mean King problem, for the understanding of the Feynman path integral formalism and for the studies of the Weyl-Heisenberg group (in connection with quantum optics). MUBs also proved to be useful in classical information theory (network communication protocols) and in quantum information theory as, for instance, in quantum signal processing, quantum tomography (deciphering an unknown quantum state), quantum cryptography (secure quantum key exchange) and quantum teleportation. Along this line, measurements corresponding to MUBs are appropriate for an optimal determination of the density matrix of a quantum system, and the use of MUBs ensures maximum security for quantum communication (the Bohr principle is at the root of the BB84 quantum cryptography protocol). Let us also mention that MUBs are connected with the notion of maximal entanglement of quantum states; a result of great importance for quantum computing.

4.1.4. *Well-known results*

The main results concerning MUBs are:

1) MUBs are stable under unitary or anti-unitary transformations. More precisely, if two unbiased bases undergo the same unitary or anti-unitary transformation, they remain mutually unbiased.

2) The number $N(d)$ of MUBs in $\mathbb{C}^d$ cannot exceed $d + 1$. Thus

$$N(d) \leq d + 1$$

3) The maximum number $d + 1$ of MUBs is attained when d is a power p^m $(m \geq 1)$ of a prime number p. Thus

$$N(p^m) = p^m + 1$$

4) When d is a composite number, $N(d)$ is not known but it can be shown that

$$3 \leq N(d) \leq d + 1$$

As a more accurate result, for $d = \prod_i p_i^{m_i}$ with p_i prime and m_i positive integer, we have

$$\min(p_i^{m_i}) + 1 \leq N(d) \leq d + 1$$

By way of illustration, let us mention the following cases.

– In the particular composite case $d = 6 = 2 \times 3$, we have

$$3 \leq N(6) \leq 7$$

and it was conjectured that $N(6) = 3$. Indeed, in spite of an enormous amount of computational works, no more than three MUBs were found for $d = 6$.

– For $d = 15 = 3 \times 5$ and $d = 21 = 3 \times 7$, there are at least four MUBs.

– For $d = 676 = 2^2 \times 13^2$, we have

$$2^2 + 1 = 5 \leq N(676) \leq 677$$

but it is known how to construct at least six MUBs.

A set of $d + 1$ MUBs in $\mathbb{C}^d$ is referred to as a complete set. Such sets exist for $d = p^m$ (p prime and m positive integer) and this result opens the way to establish a link between MUBs and Galois fields and/or Galois rings.

Note that

$$d + 1 = \frac{d^2 - 1}{d - 1}$$

is the number of different measurements to fully determine a quantum state for a quantum system in dimension d. (This follows from the fact that a $d \times d$ density matrix, i.e. an Hermitian matrix with a trace equal to 1, contains $d^2 - 1$ real parameters and each measurement gives $d - 1$ real parameters.) Note also that $d^2 - 1$ and $d - 1$ are the number of generators and the rank of the special unitary group $\mathrm{SU}(d)$ in d dimensions, respectively, and that for $d = p$ (prime number) their ratio $p + 1$ is the number of disjoint sets of $p - 1$ commuting generators of $\mathrm{SU}(p)$.

For d composite, the question of knowing if there exist complete sets in dimension d, i.e. to know if $N(d)$ can be equal to $d+1$, is still an open problem (in 2017). Indeed, for d different from a power of a prime, it was conjectured (SPR conjecture) that the problem of the existence of a set of $d + 1$ MUBs in $\mathbb{C}^d$ is equivalent to the problem of whether there exist projective planes of order d.

4.2. Quantum angular momentum bases

This section is devoted to the derivation of non-standard bases for the irreducible representations of the group SU(2) from the quantum theory of angular momentum (connected to the Lie algebra of the group SU(2)). The notations are those of quantum mechanics and quantum information. In particular, $\langle\psi|\phi\rangle$ stands for the inner product of the vector $|\phi\rangle$ of $\mathbb{C}^d$ by the vector $|\psi\rangle$ of $\mathbb{C}^d$.

4.2.1. *Standard basis for SU*(2)

In quantum information, we use qubits which are, indeed, normalized vectors in the Hilbert space $\mathbb{C}^2$ of dimension $d = 2$. The more general qubit

$$|\psi_2\rangle = c_0|0\rangle + c_1|1\rangle \text{ with } |c_0|^2 + |c_1|^2 = 1, \quad c_0 \in \mathbb{C}, \quad c_1 \in \mathbb{C}$$

is a linear combination of the vectors $|0\rangle$ and $|1\rangle$ which constitute an orthonormal (orthogonal and normalized) basis

$$B_2 = \{|0\rangle, |1\rangle\}$$

of $\mathbb{C}^2$. In terms of group theory, the two vectors $|0\rangle$ and $|1\rangle$ can be considered as the basis vectors for the fundamental irreducible representation $\left(\frac{1}{2}\right)$ of SU(2), in the chain SU(2) $\supset$ U(1), with

$$|0\rangle = |\frac{1}{2}, \frac{1}{2}\rangle, \quad |1\rangle = |\frac{1}{2}, -\frac{1}{2}\rangle$$

in the notations of quantum angular momentum theory.

More generally, in dimension d, we use qudits of the form

$$|\psi_d\rangle = \sum_{n=0}^{d-1} c_n|n\rangle \text{ with } \sum_{n=0}^{d-1} |c_n|^2 = 1, \quad c_n \in \mathbb{C}$$

expressed in the orthonormal basis

$$B_d = \{|n\rangle \mid n = 0, 1, \cdots, d-1\}$$

of $\mathbb{C}^d$ (with $d = 2, 3, \cdots$), the vectors of the basis B_d satisfying the orthonormality relations

$$\langle n|n'\rangle = \delta(n, n'), \quad n, n' = 0, 1, \cdots, d-1$$

In quantum information (respectively, quantum mechanics), B_d is called a computational basis (respectively, a canonical or Fock basis). Contact with the representation theory of $\mathrm{SU}(2)$ and quantum angular momentum theory can be done by introducing

$$j = \frac{1}{2}(d-1), \quad m = n - \frac{1}{2}(d-1)$$

$$|j, m\rangle = |d-1-n\rangle \Leftrightarrow |j, -m\rangle = |n\rangle$$

so that $2j = 1, 2, 3, \cdots$ and $m = j, j-1, \cdots, -j$ with the correspondence

$$|0\rangle = |j, j\rangle, \quad |1\rangle = |j, j-1\rangle, \quad \cdots, \quad |d-1\rangle = |j, -j\rangle$$

between qudits and angular momentum states. Therefore, the vectors $|n\rangle$ (with $n = 0, 1, \cdots, d-1$) of the computational basis can be viewed as the basis vectors $|j, m\rangle$ (with $m = j, j-1, \cdots, -j$) for the irreducible representation (j) of $\mathrm{SU}(2)$ in the chain $\mathrm{SU}(2) \supset \mathrm{U}(1)$. In the language of group theory and quantum angular momentum theory, the vector $|j, m\rangle$ is a common eigenvector of the Casimir operator J^2 (the square of an angular momentum) and of a Cartan generator J_z (the z component of the angular momentum) of the Lie algebra of $\mathrm{SU}(2)$. More precisely, we have the eigenvalue equations

$$J^2|j, m\rangle = j(j+1)|j, m\rangle, \quad J_z|j, m\rangle = m|j, m\rangle$$

with the orthonormality relations

$$\langle j, m|j, m'\rangle = \delta(m, m'), \quad m, m' = j, j-1, \cdots, -j$$

In other words, the computational basis B_d can be visualized as the basis

$$B_{2j+1} = \{|j, m\rangle \mid m = j, j-1, \cdots, -j\}$$

which is known as the standard basis for the irreducible representation (j) of $\mathrm{SU}(2)$ or the angular momentum basis corresponding to the angular momentum quantum number j (referred to as spin angular momentum for $j = \frac{1}{2}$).

4.2.2. *Non-standard bases for SU*(2)

We are now in a position to introduce a family of non-standard bases for SU(2) which shall be connected in section 4.3 to the MUBs of quantum information. As far as the representation theory of SU(2) is concerned, we can replace the complete set $\{J^2, J_z\}$ with another complete set of two commuting operators. For instance, we may consider the set $\{J^2, v_a\}$, where the unitary operator v_a is defined by

$$v_a|j,m\rangle = \begin{cases} |j,-j\rangle \text{ if } m = j \\ \omega^{(j-m)a}|j,m+1\rangle \text{ if } m = j-1, j-2, \cdots, -j \end{cases}$$

where ω is a primitive $(2j+1)$-th root of unity, i.e.

$$\omega = \mathrm{e}^{\mathrm{i}\frac{2\pi}{2j+1}}$$

and a is a fixed parameter in the ring $\mathbb{Z}_{2j+1}$. For fixed a, the common eigenvectors of J^2 and v_a provide an alternative basis to that given by the common eigenstates of J^2 and J_z. This can be made precise by the following result.

PROPOSITION 4.1.– For fixed j and a (with $2j \in \mathbb{N}_1$ and $a \in \mathbb{Z}_{2j+1}$), the $2j+1$ common eigenvectors of J^2 and v_a can be taken in the form

$$|j\alpha;a\rangle = \frac{1}{\sqrt{2j+1}} \sum_{m=-j}^{j} \omega^{\frac{1}{2}(j+m)(j-m+1)a+(j+m)\alpha}|j,m\rangle$$

with $\alpha = 0, 1, \cdots, 2j$. The corresponding eigenvalues of v_a are given by

$$v_a|j\alpha;a\rangle = \omega^{ja-\alpha}|j\alpha;a\rangle$$

Then, the spectrum of v_a is nondegenerate.

The inner product

$$\langle j\alpha;a|j\beta;a\rangle = \delta(\alpha,\beta), \quad \alpha,\beta = 0,1,\cdots,2j$$

shows that for fixed j and a

$$B_a = \{|j\alpha; a\rangle \mid \alpha = 0, 1, \cdots, 2j\}$$

is an orthonormal set that provides a non-standard basis for the irreducible representation (j) of SU(2). For fixed j, there exists $2j+1$ orthonormal bases B_a, since a can take $2j+1$ distinct values ($a = 0, 1, \cdots, 2j$).

4.2.3. *Bases in quantum information*

We now go back to quantum information. By introducing the change of notations

$$d = 2j+1, \quad n = j+m, \quad |n\rangle = |j, -m\rangle, \quad |a\alpha\rangle = |j\alpha; a\rangle$$

adapted to quantum information, the eigenvectors of v_a can be written as

$$\begin{aligned} |a\alpha\rangle &= \frac{1}{\sqrt{d}} \sum_{n \in \mathbb{Z}_d} \omega^{\frac{1}{2}n(d-n)a+n\alpha} |d-1-n\rangle \\ &= \frac{1}{\sqrt{d}} \sum_{n \in \mathbb{Z}_d} \omega^{\frac{1}{2}(n+1)(d-n-1)a-(n+1)\alpha} |n\rangle \end{aligned}$$

where $\omega = \mathrm{e}^{\mathrm{i}\frac{2\pi}{d}}$. The vector $|a\alpha\rangle$ satisfies the eigenvalue equation

$$v_a |a\alpha\rangle = \omega^{\frac{1}{2}(d-1)a-\alpha} |a\alpha\rangle$$

For fixed d and a, each eigenvector $|a\alpha\rangle$ is a linear combination of the qudits $|0\rangle, |1\rangle, \cdots, |d-1\rangle$ and the basis

$$B_a = \{|a\alpha\rangle \mid \alpha = 0, 1, \cdots, d-1\}$$

is an alternative to the computational basis B_d. For fixed d, we therefore have $d+1$ remarkable bases of the d-dimensional space $\mathbb{C}^d$, namely, B_d and B_a for $a = 0, 1, \cdots, d-1$.

The operator v_a can be represented by a d-dimensional unitary matrix V_a. The matrix V_a, built on the basis B_d with the ordering $0, 1, \cdots, d-1$ for the lines and columns, reads

$$V_a = \begin{pmatrix} 0 & \omega^a & 0 & \cdots & 0 \\ 0 & 0 & \omega^{2a} & \cdots & 0 \\ \vdots & \vdots & \vdots & \cdots & \vdots \\ 0 & 0 & 0 & \cdots & \omega^{(d-1)a} \\ 1 & 0 & 0 & \cdots & 0 \end{pmatrix}$$

The eigenvectors of V_a are

$$\begin{aligned} \phi(a\alpha) &= \frac{1}{\sqrt{d}} \sum_{n \in \mathbb{Z}_d} \omega^{\frac{1}{2}n(d-n)a+n\alpha} \phi_{d-1-n} \\ &= \frac{1}{\sqrt{d}} \sum_{n \in \mathbb{Z}_d} \omega^{\frac{1}{2}(n+1)(d-n-1)a-(n+1)\alpha} \phi_n \end{aligned}$$

with $\alpha = 0, 1, \cdots, d-1$, where ϕ_n with $n = 0, 1, \cdots, d-1$ are the column vectors

$$\phi_0 = \begin{pmatrix} 1 \\ 0 \\ \vdots \\ 0 \end{pmatrix}, \quad \phi_1 = \begin{pmatrix} 0 \\ 1 \\ \vdots \\ 0 \end{pmatrix}, \quad \cdots, \quad \phi_{d-1} = \begin{pmatrix} 0 \\ 0 \\ \vdots \\ 1 \end{pmatrix}$$

representing the qudits $|0\rangle, |1\rangle, \cdots, |d-1\rangle$, respectively. The vectors $\phi(a\alpha)$ satisfy the eigenvalue equation

$$V_a \phi(a\alpha) = \omega^{\frac{1}{2}(d-1)a-\alpha} \phi(a\alpha)$$

with the orthonormality relation

$$\phi(a\alpha)^\dagger \phi(a\beta) = \delta(\alpha, \beta)$$

for $\alpha, \beta = 0, 1, \cdots, d-1$.

The matrix V_a can be diagonalized by means of the d-dimensional matrix H_a of elements

$$(H_a)_{n\alpha} = \frac{1}{\sqrt{d}}\omega^{\frac{1}{2}(n+1)(d-n-1)a-(n+1)\alpha}$$

with the lines and columns of H_a arranged from left to right and from top to bottom in the order $n, \alpha = 0, 1, \cdots, d-1$. Indeed, by introducing the $d \times d$ permutation matrix

$$P = \begin{pmatrix} 1 & 0 & 0 & \cdots & 0 & 0 \\ 0 & 0 & 0 & \cdots & 0 & 1 \\ 0 & 0 & 0 & \cdots & 1 & 0 \\ \vdots & \vdots & \vdots & \cdots & \vdots & \vdots \\ 0 & 0 & 1 & \cdots & 0 & 0 \\ 0 & 1 & 0 & \cdots & 0 & 0 \end{pmatrix}$$

we can check that

$$(H_aP)^\dagger V_a (H_aP) = \omega^{\frac{1}{2}(d-1)a} \begin{pmatrix} \omega^0 & 0 & \cdots & 0 \\ 0 & \omega^1 & \cdots & 0 \\ \vdots & \vdots & \cdots & \vdots \\ 0 & 0 & \cdots & \omega^{d-1} \end{pmatrix}$$

from which we recover the eigenvalues of V_a. Note that the complex matrix H_a is a unitary matrix for which each entry has a modulus equal to $\frac{1}{\sqrt{d}}$. Thus, H_a is a generalized Hadamard matrix.

As an illustration, we give below the eigenvectors $\phi(a\alpha)$ of V_a for $d = 2$ and 3. They can be transcribed in terms of the eigenvectors $|a\alpha\rangle$ of the operator v_a owing to the replacements $\phi(a\alpha) \to |a\alpha\rangle$ and $\phi_n \to |n\rangle$.

– For $d = 2$, we have the two bases B_0 and B_1 (a can take the values 0 and 1). The matrix

$$V_a = \begin{pmatrix} 0 & \omega^a \\ 1 & 0 \end{pmatrix}, \quad \omega = \mathrm{e}^{i\pi}$$

has the eigenvectors

$$\phi(a\alpha) = \frac{1}{\sqrt{2}}(\omega^{\frac{1}{2}a+\alpha}\phi_0 + \phi_1), \quad \alpha = 0, 1$$

This leads to the bases

$$B_0 : \phi(00) = \frac{1}{\sqrt{2}}(\phi_0 + \phi_1), \quad \phi(01) = -\frac{1}{\sqrt{2}}(\phi_0 - \phi_1)$$

$$B_1 : \phi(10) = \frac{\mathrm{i}}{\sqrt{2}}(\phi_0 - \mathrm{i}\phi_1), \quad \phi(11) = -\frac{\mathrm{i}}{\sqrt{2}}(\phi_0 + \mathrm{i}\phi_1)$$

The bases B_0 and B_1 together with the computational basis B_2 are familiar bases for representing qubits (up to an interchange $\phi_0 \leftrightarrow \phi_1$ for B_0 and B_1, compare with example 4.1.2).

– For $d = 3$, we have the three bases B_0, B_1 and B_2 (since a can be 0, 1 and 2). In this case, the matrix

$$V_a = \begin{pmatrix} 0 & \omega^a & 0 \\ 0 & 0 & \omega^{2a} \\ 1 & 0 & 0 \end{pmatrix}, \quad \omega = \mathrm{e}^{\mathrm{i}\frac{2\pi}{3}}$$

admits the eigenvectors

$$\phi(a\alpha) = \frac{1}{\sqrt{3}}\left(\omega^{a+2\alpha}\phi_0 + \omega^{a+\alpha}\phi_1 + \phi_2\right), \quad \alpha = 0, 1, 2$$

This yields the bases

$$B_0 : \phi(00) = \frac{1}{\sqrt{3}}(\phi_0 + \phi_1 + \phi_2)$$

$$\phi(01) = \frac{1}{\sqrt{3}}\left(\omega^2\phi_0 + \omega\phi_1 + \phi_2\right)$$

$$\phi(02) = \frac{1}{\sqrt{3}}\left(\omega\phi_0 + \omega^2\phi_1 + \phi_2\right)$$

$$B_1 : \phi(10) = \frac{1}{\sqrt{3}}(\omega\phi_0 + \omega\phi_1 + \phi_2)$$

$$\phi(11) = \frac{1}{\sqrt{3}}(\phi_0 + \omega^2\phi_1 + \phi_2)$$

$$\phi(12) = \frac{1}{\sqrt{3}}(\omega^2\phi_0 + \phi_1 + \phi_2)$$

$$B_2 : \phi(20) = \frac{1}{\sqrt{3}}(\omega^2\phi_0 + \omega^2\phi_1 + \phi_2)$$

$$\phi(21) = \frac{1}{\sqrt{3}}(\omega\phi_0 + \phi_1 + \phi_2)$$

$$\phi(22) = \frac{1}{\sqrt{3}}(\phi_0 + \omega\phi_1 + \phi_2)$$

which, together with the computational basis B_3, are of interest for the so-called qutrits of quantum information.

4.3. SU(2) approach to mutually unbiased bases

4.3.1. *A master formula for $d = p$ (p prime)*

Going back to the case where d is arbitrary, we now examine an important property for the couple (B_a, B_d) and its generalization to couples (B_a, B_b) with $b \neq a$ $(a, b = 0, 1, \cdots, d-1)$. For fixed d and a, we verify that

$$|\langle n|a\alpha\rangle| = \frac{1}{\sqrt{d}}, \quad n, \alpha = 0, 1, \cdots, d-1$$

which shows that B_a and B_d are two unbiased bases of the Hilbert space $\mathbb{C}^d$.

Other examples of unbiased bases can be obtained for $d = 2$ and 3. We easily check that the bases B_0 and B_1, for $d = 2$, are unbiased. Similarly, the bases B_0, B_1 and B_2, for $d = 3$, are mutually unbiased. Therefore, by taking into account the computational basis B_d, we end up with $d + 1 = 3$ MUBs for $d = 2$ and $d + 1 = 4$ MUBs for $d = 3$. This is in agreement with the general result according to which, in dimension d, the maximum number $d + 1$ of MUBs is attained when d is a prime number or a power of a prime number. The results for $d = 2$ and 3 can be generalized through the following proposition.

PROPOSITION 4.2.– For $d = p$, with p a prime number, the bases $B_0, B_1, \cdots, B_p$ form a complete set of $p+1$ MUBs. The p^2 vectors $|a\alpha\rangle$, with $a, \alpha = 0, 1, \cdots, p-1$, of the bases $B_0, B_1, \cdots, B_{p-1}$ are given by a single formula, namely

$$|a\alpha\rangle = \frac{1}{\sqrt{p}} \sum_{n \in \mathbb{F}_p} \omega^{\frac{1}{2}n(p-n)a + n\alpha} |p-1-n\rangle$$

$$= \frac{1}{\sqrt{p}} \sum_{n \in \mathbb{F}_p} \omega^{\frac{1}{2}(n+1)(p-n-1)a - (n+1)\alpha} |n\rangle, \quad \omega = \mathrm{e}^{\mathrm{i}\frac{2\pi}{p}}$$

which gives the p basis vectors for each basis B_a. In the matrix form, $|a\alpha\rangle$ and $|n\rangle$ are replaced by $\phi(a\alpha)$ and ϕ_n, respectively.

PROOF.– First, the computational basis B_p is clearly unbiased to any of the p bases $B_0, B_1, \cdots, B_{p-1}$. Second, let us consider

$$\langle a\alpha|b\beta\rangle = \frac{1}{p} \sum_{k=0}^{p-1} \omega^{\frac{1}{2}k(p-k)(b-a) + k(\beta-\alpha)}$$

$$= \frac{1}{p} \sum_{k=0}^{p-1} \mathrm{e}^{\mathrm{i}\frac{\pi}{p}\{(a-b)k^2 + [(b-a)p + 2(\beta-\alpha)]k\}}$$

for $b \neq a$. The inner product $\langle a\alpha|b\beta\rangle$ can be rewritten by making use of the generalized quadratic Gauss sum (see Chapter 5)

$$S(u, v, w) = \sum_{k=0}^{|w|-1} \mathrm{e}^{\mathrm{i}\frac{\pi}{w}(uk^2 + vk)}$$

where u, v and w are the integers such that u and w are co-prime, uw is non-vanishing and $uw + v$ is even. This leads to

$$\langle a\alpha|b\beta\rangle = \frac{1}{p} S(u, v, w)$$

with

$$u = a - b, \quad v = -(a-b)p - 2(\alpha - \beta), \quad w = p$$

From section 5.1.7 of Chapter 5, it is possible to show that $|S(u, v, w)| = \sqrt{p}$. This leads to

$$|\langle a\alpha|b\beta\rangle| = \frac{1}{\sqrt{p}}$$

for $b \neq a$ and $\alpha, \beta = 0, 1, \cdots, p-1$. This completes the proof. □

It is not necessary to treat the cases p odd and p even separately: the master formula for $|a\alpha\rangle$ given in the above proposition is valid both for p even prime ($p = 2$) and for p odd prime.

In many of the papers dealing with the construction of MUBs for $d = p$ a prime number or $d = p^m$ a power of a prime number, the explicit derivation of the bases requires the diagonalization of a set of matrices. The master formula arises from the diagonalization of a single matrix. It allows us to derive in one step the $(p+1)p$ vectors (or qupits, i.e. qudits with $d = p$) of a complete set of $p+1$ MUBs in $\mathbb{C}^p$ via a single formula easily codable on a classical computer.

Note that, for d arbitrary, the inner product $\langle a\alpha|b\beta\rangle$ can be rewritten as

$$\langle a\alpha|b\beta\rangle = \left(H_a{}^{\dagger} H_b\right)_{\alpha\beta}$$

in terms of the generalized Hadamard matrices H_a and H_b. In the case where $d = p$ is a prime number, we find that

$$\left|\left(H_a{}^{\dagger} H_b\right)_{\alpha\beta}\right| = |\langle a\alpha|b\beta\rangle| = \frac{1}{\sqrt{p}}$$

Therefore, the product $H_a{}^{\dagger} H_b$ is another generalized Hadamard matrix.

Finally, note that the passage, given by the master formula, from the computational basis $B_p = \{|n\rangle \mid n = 0, 1, \cdots, p-1\}$ to the basis $B_0 = \{|0\alpha\rangle \mid \alpha = 0, 1, \cdots, p-1\}$ corresponds to a discrete Fourier transform. Similarly, the passage from the basis B_p to the basis $B_a = \{|a\alpha\rangle \mid \alpha = 0, 1, \cdots, p-1\}$ with $a = 1, 2, \cdots, p-1$ corresponds to a *discrete quadratic Fourier transform*.

4.3.2. *Examples:* $d = 2$ *and* 3

4.3.2.1. *The case* $d = 2$

In this case, relevant for a spin $j = \frac{1}{2}$ or for a qubit, we have $\omega = \exp(\mathrm{i}\pi)$ and $a, \alpha \in \mathbb{F}_2$. The matrices of the operators v_a are

$$V_0 = \begin{pmatrix} 0 & 1 \\ 1 & 0 \end{pmatrix} = \sigma_x, \quad V_1 = \begin{pmatrix} 0 & -1 \\ 1 & 0 \end{pmatrix} = -\mathrm{i}\sigma_y$$

where σ_x and σ_y are two Pauli matrices. The $d+1 = 3$ MUBs B_0, B_1 and B_2 are the following

$$B_2 : \ |0\rangle, \quad |1\rangle$$

$$B_0 : \ |00\rangle = \frac{1}{\sqrt{2}}\left(|0\rangle + |1\rangle\right), \quad |01\rangle = -\frac{1}{\sqrt{2}}\left(|0\rangle - |1\rangle\right)$$

$$B_1 : \ |10\rangle = \frac{\mathrm{i}}{\sqrt{2}}\left(|0\rangle - \mathrm{i}|1\rangle\right), \quad |11\rangle = -\frac{\mathrm{i}}{\sqrt{2}}\left(|0\rangle + \mathrm{i}|1\rangle\right)$$

In terms of eigenvectors of the matrices V_a, we must replace the state vector $|a\alpha\rangle$ by the column vector $\phi(a\alpha)$. This leads to

$$B_2 : \ |0\rangle \to \begin{pmatrix} 1 \\ 0 \end{pmatrix}, \quad |1\rangle \to \begin{pmatrix} 0 \\ 1 \end{pmatrix}$$

$$B_0 : \ |00\rangle \to \frac{1}{\sqrt{2}}\begin{pmatrix} 1 \\ 1 \end{pmatrix}, \quad |01\rangle \to -\frac{1}{\sqrt{2}}\begin{pmatrix} 1 \\ -1 \end{pmatrix}$$

$$B_1 : \ |10\rangle \to \frac{\mathrm{i}}{\sqrt{2}}\begin{pmatrix} 1 \\ -\mathrm{i} \end{pmatrix}, \quad |11\rangle \to -\frac{\mathrm{i}}{\sqrt{2}}\begin{pmatrix} 1 \\ \mathrm{i} \end{pmatrix}$$

4.3.2.2. *The case* $d = 3$

This case corresponds to an angular momentum $j = 1$ or to a qutrit. Here, we have $\omega = \exp(\mathrm{i}\frac{2\pi}{3})$ and $a, \alpha \in \mathbb{F}_3$. The matrices of the operators v_a are

$$V_0 = \begin{pmatrix} 0 & 1 & 0 \\ 0 & 0 & 1 \\ 1 & 0 & 0 \end{pmatrix}, \quad V_1 = \begin{pmatrix} 0 & \omega & 0 \\ 0 & 0 & \omega^2 \\ 1 & 0 & 0 \end{pmatrix}, \quad V_2 = \begin{pmatrix} 0 & \omega^2 & 0 \\ 0 & 0 & \omega \\ 1 & 0 & 0 \end{pmatrix}$$

The $d+1=4$ MUBs B_3, B_0, B_1 and B_2 are the following.

$$
\begin{aligned}
B_3 :\ & |0\rangle, \quad |1\rangle, \quad |2\rangle \\
B_0 :\ & |00\rangle = \frac{1}{\sqrt{3}}\left(|0\rangle + |1\rangle + |2\rangle\right) \\
& |01\rangle = \frac{1}{\sqrt{3}}\left(\omega^2|0\rangle + \omega|1\rangle + |2\rangle\right) \\
& |02\rangle = \frac{1}{\sqrt{3}}\left(\omega|0\rangle + \omega^2|1\rangle + |2\rangle\right) \\
B_1 :\ & |10\rangle = \frac{1}{\sqrt{3}}\left(\omega|0\rangle + \omega|1\rangle + |2\rangle\right) \\
& |11\rangle = \frac{1}{\sqrt{3}}\left(|0\rangle + \omega^2|1\rangle + |2\rangle\right) \\
& |12\rangle = \frac{1}{\sqrt{3}}\left(\omega^2|0\rangle + |1\rangle + |2\rangle\right) \\
B_2 :\ & |20\rangle = \frac{1}{\sqrt{3}}\left(\omega^2|0\rangle + \omega^2|1\rangle + |2\rangle\right) \\
& |21\rangle = \frac{1}{\sqrt{3}}\left(\omega|0\rangle + |1\rangle + |2\rangle\right) \\
& |22\rangle = \frac{1}{\sqrt{3}}\left(|0\rangle + \omega|1\rangle + |2\rangle\right)
\end{aligned}
$$

This can be transcribed in terms of column vectors as follows

$$
B_3 :\ |0\rangle \to \begin{pmatrix} 1 \\ 0 \\ 0 \end{pmatrix}, \quad |1\rangle \to \begin{pmatrix} 0 \\ 1 \\ 0 \end{pmatrix}, \quad |2\rangle \to \begin{pmatrix} 0 \\ 0 \\ 1 \end{pmatrix}
$$

$$
B_0 :\ |00\rangle \to \frac{1}{\sqrt{3}}\begin{pmatrix} 1 \\ 1 \\ 1 \end{pmatrix}, \ |01\rangle \to \frac{1}{\sqrt{3}}\begin{pmatrix} \omega^2 \\ \omega \\ 1 \end{pmatrix}, \ |02\rangle \to \frac{1}{\sqrt{3}}\begin{pmatrix} \omega \\ \omega^2 \\ 1 \end{pmatrix}
$$

$$
B_1 :\ |10\rangle \to \frac{1}{\sqrt{3}}\begin{pmatrix} \omega \\ \omega \\ 1 \end{pmatrix}, \ |11\rangle \to \frac{1}{\sqrt{3}}\begin{pmatrix} 1 \\ \omega^2 \\ 1 \end{pmatrix}, \ |12\rangle \to \frac{1}{\sqrt{3}}\begin{pmatrix} \omega^2 \\ 1 \\ 1 \end{pmatrix}
$$

$$B_2 : \ |20\rangle \to \frac{1}{\sqrt{3}} \begin{pmatrix} \omega^2 \\ \omega^2 \\ 1 \end{pmatrix}, \ |21\rangle \to \frac{1}{\sqrt{3}} \begin{pmatrix} \omega \\ 1 \\ 1 \end{pmatrix}, \ |22\rangle \to \frac{1}{\sqrt{3}} \begin{pmatrix} 1 \\ \omega \\ 1 \end{pmatrix}$$

4.3.3. *An alternative formula for* $d = p$ *(*p *odd prime)*

In the special case where $d = p$ is an odd prime number, the formula

$$|a\alpha\rangle' = \frac{1}{\sqrt{p}} \sum_{n \in \mathbb{F}_p} \omega^{(an+\alpha)n} |n\rangle, \quad \omega = \mathrm{e}^{\mathrm{i}\frac{2\pi}{p}}$$

provides an alternative to the master formula given in section 4.3.1. Indeed, it can be shown that

$$B_a{}' = \{|a\alpha\rangle' \mid \alpha = 0, 1, \cdots, p-1\}$$

where a can take any of the values $0, 1, \cdots, p-1$ constitutes an orthonormal basis of $\mathbb{C}^d$ and that the p bases $B_a{}'$ ($a = 0, 1, \cdots, p-1$) form, with the computational basis B_p, a complete set of $p+1$ MUBs. The proof, based on the properties of Gauss sums, is analogous to that given in section 4.3.1.

It is to be emphasized that for p even prime ($p = 2$), the bases $B_0{}'$, $B_1{}'$ and B_2 do not form a complete set of MUBs whereas the proposition given in section 4.3.1 is valid for p odd prime and equally well for p even prime.

4.3.4. *Weyl pairs*

4.3.4.1. *Shift and phase operators*

Let us go back to the case d arbitrary. The matrix V_a can be decomposed as

$$V_a = XZ^a, \quad a = 0, 1, \cdots, d-1$$

where

$$X = \begin{pmatrix} 0 & 1 & 0 & \cdots & 0 \\ 0 & 0 & 1 & \cdots & 0 \\ \vdots & \vdots & \vdots & \cdots & \vdots \\ 0 & 0 & 0 & \cdots & 1 \\ 1 & 0 & 0 & \cdots & 0 \end{pmatrix}, \quad Z = \begin{pmatrix} 1 & 0 & 0 & \cdots & 0 \\ 0 & \omega & 0 & \cdots & 0 \\ 0 & 0 & \omega^2 & \cdots & 0 \\ \vdots & \vdots & \vdots & \cdots & \vdots \\ 0 & 0 & 0 & \cdots & \omega^{d-1} \end{pmatrix}$$

$$\omega = \mathrm{e}^{\mathrm{i}\frac{2\pi}{d}}.$$

The matrices X and Z satisfy

$$Z\phi_n = \omega^n \phi_n, \quad n = 0, 1, \cdots, d-1$$

$$X\phi_n = \phi_{n-1 \bmod d} = \begin{cases} \phi_{n-1}, & n = 1, 2, \cdots, d-1 \\ \phi_{d-1}, & n = 0 \end{cases}$$

The linear operators corresponding to the matrices X and Z are known in quantum information as flip or shift and clock or phase operators, respectively. The unitary matrices X and Z ω-commute in the sense that

$$XZ - \omega ZX = O_d$$

In addition, they satisfy

$$X^d = Z^d = I_d$$

where I_d and O_d are the d-dimensional unity and zero matrices, respectively. The last two equations show that X and Z constitute the so-called Weyl pair.

Note that the Weyl pair (X, Z) can be deduced from the master matrix V_a via

$$X = V_0, \quad Z = V_0{}^{\dagger} V_1$$

which shows a further interest of the matrix V_a. Indeed, the matrix V_a condensates all that can be done with the matrices X and Z. This has been seen in section 4.3.1 with the derivation of a single formula for the determination from V_a of a complete set of $p+1$ MUBs when $d = p$ is a prime, whereas many other determinations of such a complete set needs repeated use of the matrices X and Z.

A connection between X and Z can be deduced from the expression of $(H_a P)^{\dagger} V_a (H_a P)$ given in 4.2.3. By taking $a = 0$, we obtain

$$(H_0 P)^{\dagger} X (H_0 P) = Z \Leftrightarrow X = (H_0 P) Z (H_0 P)^{\dagger}$$

where H_0 is the matrix of a discrete Fourier transform that allows passing from the vectors ϕ_n $(n = 0, 1, \cdots, d-1)$ to the vector $\phi(0, \alpha)$, according to

$$\phi(0,\alpha) = \sum_{n\in\mathbb{Z}_d} (H_0)_{n\alpha}\,\phi_n = (-1)^{\alpha}\frac{1}{\sqrt{d}}\sum_{n\in\mathbb{Z}_d} \mathrm{e}^{-\mathrm{i}\frac{2\pi}{d}n\alpha}\phi_n$$

see the expression of $\phi(a, \alpha)$ in 4.2.3.

4.3.4.2. *Generalized Pauli matrices*

For d arbitrary, let us define the matrices

$$U_{ab} = X^a Z^b, \quad a, b \in \mathbb{Z}_d$$

The matrices U_{ab} belong to the unitary group $\mathrm{U}(d)$. The d^2 matrices U_{ab} are called *generalized Pauli matrices* in dimension d. They satisfy the trace relation

$$\mathrm{tr}\left({U_{ab}}^{\dagger} U_{a'b'}\right) = d\,\delta(a,a')\,\delta(b,b')$$

Thus, the set $\{U_{ab} \mid a, b \in \mathbb{Z}_d\}$ of unitary matrices is an orthogonal set with respect to the Hilbert-Schmidt inner product. Consequently, the d^2 pairwise orthogonal matrices U_{ab} can be used as a basis of $\mathbb{C}^{d\times d}$.

EXAMPLE 4.1.– The case $d = 2 \Leftrightarrow j = \frac{1}{2}$ ($\Rightarrow \omega = \exp(\mathrm{i}\pi)$ and $a, b = 0, 1$) corresponds to the two-dimensional ordinary Pauli matrices of quantum mechanics. The matrices $X^a Z^b$ are

$$I_2 = X^0 Z^0 = \begin{pmatrix} 1 & 0 \\ 0 & 1 \end{pmatrix}, \quad X = X^1 Z^0 = \begin{pmatrix} 0 & 1 \\ 1 & 0 \end{pmatrix}$$

$$Z = X^0 Z^1 = \begin{pmatrix} 1 & 0 \\ 0 & -1 \end{pmatrix}, \quad Y = X^1 Z^1 = \begin{pmatrix} 0 & -1 \\ 1 & 0 \end{pmatrix}$$

so that the matrices X and Z generate the ordinary Pauli matrices. Indeed, in terms of the usual (Hermitian and unitary) Pauli matrices σ_0, σ_x, σ_y and σ_z, we have

$$I_2 = \sigma_0, \quad X = V_0 = \sigma_x, \quad Y = XZ = V_1 = -\mathrm{i}\sigma_y, \quad Z = \sigma_z$$

where

$$\sigma_0 = \begin{pmatrix} 1 & 0 \\ 0 & 1 \end{pmatrix}, \quad \sigma_x = \begin{pmatrix} 0 & 1 \\ 1 & 0 \end{pmatrix}, \quad \sigma_y = \begin{pmatrix} 0 & -\mathrm{i} \\ \mathrm{i} & 0 \end{pmatrix}, \quad \sigma_z = \begin{pmatrix} 1 & 0 \\ 0 & -1 \end{pmatrix}$$

EXAMPLE 4.2.– The case $d = 3 \Leftrightarrow j = 1$ ($\Rightarrow \omega = \exp(\mathrm{i}\frac{2\pi}{3})$ and $a, b = 0, 1, 2$) yields nine three-dimensional matrices. More precisely, the matrices X and Z generate $I_3 = X^0 Z^0$ and

$$X = V_0, \quad X^2, \quad Z, \quad Z^2, \quad XZ = V_1, \quad X^2Z^2, \quad XZ^2 = V_2, \quad X^2Z$$

In detail, the matrices $X^a Z^b$ are

$$X^0Z^0 = \begin{pmatrix} 1 & 0 & 0 \\ 0 & 1 & 0 \\ 0 & 0 & 1 \end{pmatrix}, \; X^0Z^1 = \begin{pmatrix} 1 & 0 & 0 \\ 0 & \omega & 0 \\ 0 & 0 & \omega^2 \end{pmatrix}, \; X^0Z^2 = \begin{pmatrix} 1 & 0 & 0 \\ 0 & \omega^2 & 0 \\ 0 & 0 & \omega \end{pmatrix}$$

$$X^1Z^0 = \begin{pmatrix} 0 & 1 & 0 \\ 0 & 0 & 1 \\ 1 & 0 & 0 \end{pmatrix}, \; X^1Z^1 = \begin{pmatrix} 0 & \omega & 0 \\ 0 & 0 & \omega^2 \\ 1 & 0 & 0 \end{pmatrix}, \; X^1Z^2 = \begin{pmatrix} 0 & \omega^2 & 0 \\ 0 & 0 & \omega \\ 1 & 0 & 0 \end{pmatrix}$$

$$X^2Z^0 = \begin{pmatrix} 0 & 0 & 1 \\ 1 & 0 & 0 \\ 0 & 1 & 0 \end{pmatrix}, \; X^2Z^1 = \begin{pmatrix} 0 & 0 & \omega^2 \\ 1 & 0 & 0 \\ 0 & \omega & 0 \end{pmatrix}, \; X^2Z^2 = \begin{pmatrix} 0 & 0 & \omega \\ 1 & 0 & 0 \\ 0 & \omega^2 & 0 \end{pmatrix}$$

They differ from the Gell-Mann matrices, well-known in elementary particle physics, and constitute a natural extension in dimension $d = 3$ of the usual Pauli matrices.

4.3.4.3. *Weyl pair and groups*

For arbitrary d, the Weyl pair ($X = V_0, Z = V_0^\dagger V_1$) is a basic ingredient for generating the Pauli group P_d in d dimensions and the Lie algebra of the linear group $\mathrm{GL}(d, \mathbb{C})$ in d dimensions, groups of central interest in group theory, quantum mechanics and quantum information.

The Pauli group. For arbitrary d, let us define the matrices

$$V_{abc} = \omega^a U_{bc} = \omega^a X^b Z^c, \quad a, b, c \in \mathbb{Z}_d, \quad \omega = \mathrm{e}^{\mathrm{i}\frac{2\pi}{d}}$$

The matrices V_{abc} are unitary and satisfy

$$\mathrm{tr}\left({V_{abc}}^{\dagger}V_{a'b'c'}\right) = \omega^{a'-a}\, d\, \delta(b,b')\, \delta(c,c')$$

In addition, we have the following result.

PROPOSITION 4.3.– The set $\{V_{abc} \mid a,b,c \in \mathbb{Z}_d\}$ is a finite group of order d^3, denoted as P_d, for the internal law (matrix multiplication)

$$V_{abc}V_{a'b'c'} = V_{a''b''c''}, \quad a'' = a + a' - cb', \quad b'' = b + b', \quad c'' = c + c'$$

It is a non-commutative (for $d \geq 2$) nilpotent group with a nilpotency class equal to 3.

The group P_d is called the Pauli group in dimension d. It is of considerable importance in quantum information, especially for quantum computation and for quantum correcting codes. The group P_d is a sub-group of the unitary group $\mathrm{U}(d)$. The normalizer of P_d in $\mathrm{U}(d)$ is a Clifford-type group in d dimensions denoted as C_d. More precisely, C_d is the set $\{U \in \mathrm{U}(d)$ $|UP_dU^{\dagger} = P_d\}$ endowed with matrix multiplication. The Pauli group P_d as well as any other invariant sub-group of C_d can be used for stabilizing errors in quantum computing as, for instance, in the case of N-qubit systems (corresponding to $d = 2^N$).

Moreover, the Pauli group is connected to the Heisenberg-Weyl group. In fact, the group P_d corresponds to a discretization of the Heisenberg-Weyl group $HW(\mathbb{R})$. From an abstract point of view, the group $HW(\mathbb{R})$ is the set $S = \{(x,y,z) \mid x,y,z \in \mathbb{R}\}$ equipped with the internal law $S \times S \to S$ defined via

$$(x,y,z)(x',y',z') = (x + x' - zy', y + y', z + z')$$

This group is a non-commutative Lie group of order 3. It is non-compact and nilpotent with a nilpotency class equal to 3. The passage from $HW(\mathbb{R})$ to P_d amounts to replacing the infinite field $\mathbb{R}$ with the finite ring $\mathbb{Z}_d$, so that $HW(\mathbb{R})$ gives $HW(\mathbb{Z}_d) = P_d$.

The three generators of $HW(\mathbb{R})$ are

$$H = \frac{1}{\mathrm{i}}\frac{\partial}{\partial x}, \quad Q = \frac{1}{\mathrm{i}}\frac{\partial}{\partial y}, \quad P = \frac{1}{\mathrm{i}}\left(\frac{\partial}{\partial z} - y\frac{\partial}{\partial x}\right)$$

They satisfy

$$[Q, P]_- = \mathrm{i}H, \quad [P, H]_- = 0, \quad [H, Q]_- = 0$$

Therefore, the Lie algebra $hw(\mathbb{R})$ of $HW(\mathbb{R})$ is a three-dimensional nilpotent Lie algebra with a nilpotency class equal to 3. The commutation relations of Q, P and H are reminiscent of the Heisenberg commutation relations. As a matter of fact, the Heisenberg commutation relations correspond to an infinite-dimensional irreducible representation by Hermitian matrices of $hw(\mathbb{R})$. The Lie algebra $hw(\mathbb{R})$ also admits finite-dimensional irreducible representations at the price to abandon the Hermitian character of the representation matrices.

The linear group. The Weyl pair consisting of the generalized Pauli matrices X and Z in d dimensions can be used for constructing a basis of the Lie algebra of $\mathrm{U}(d)$. More precisely, we have the two following propositions.

PROPOSITION 4.4.– For arbitrary d, the set $\{X^a Z^b \mid a, b \in \mathbb{Z}_d\}$ forms a basis for the Lie algebra $\mathrm{gl}(d, \mathbb{C})$ of the linear group $\mathrm{GL}(d, \mathbb{C})$ or for the Lie algebra $\mathrm{u}(d)$ of the unitary group $\mathrm{U}(d)$. The Lie brackets of $\mathrm{gl}(d, \mathbb{C})$ in such a basis are

$$[X^a Z^b, X^e Z^f]_- = \sum_{i \in \mathbb{Z}_d} \sum_{j \in \mathbb{Z}_d} (ab, ef; ij) X^i Z^j$$

with the structure constants

$$(ab, ef; ij) = \delta(i, a+e)\delta(j, b+f)\left(\omega^{-be} - \omega^{-af}\right)$$

where $a, b, e, f, i, j \in \mathbb{Z}_d$.

Note that the commutator $[U_{ab}, U_{ef}]_-$ and the anticommutator $[U_{ab}, U_{ef}]_+$ of U_{ab} and U_{ef} are given by

$$[U_{ab}, U_{ef}]_\pm = \left(\omega^{-be} \pm \omega^{-af}\right) U_{ij}, \quad i = a+e, \quad j = b+f$$

Consequently, $[U_{ab}, U_{ef}]_- = 0$ if and only if $af - be = 0$ (mod d) and $[U_{ab}, U_{ef}]_+ = 0$ if and only if $af - be = \frac{1}{2}d$ (mod d). Therefore, all anticommutators $[U_{ab}, U_{ef}]_+$ are different from 0 if d is an odd integer.

PROPOSITION 4.5.– For $d = p$, with p a prime number, the simple Lie algebra $\mathrm{sl}(p, \mathbb{C})$ of the special linear group $\mathrm{SL}(p, \mathbb{C})$ or its compact real form $\mathrm{su}(d)$ of the special unitary group $\mathrm{SU}(d)$ can be decomposed into a sum of $p+1$ Abelian subalgebras of dimension $p-1$

$$\mathrm{sl}(p, \mathbb{C}) = \mathcal{V}_0 \oplus \mathcal{V}_1 \oplus \cdots \oplus \mathcal{V}_p$$

where each of the $p+1$ subalgebras $\mathcal{V}_0, \mathcal{V}_1, \cdots, \mathcal{V}_p$ is a Cartan subalgebra generated by a set of $p-1$ commuting matrices.

The decomposition of $\mathrm{sl}(p, \mathbb{C})$ (called orthogonal decomposition of $\mathrm{sl}(p, \mathbb{C})$) is trivial for $p = 2$. In fact, for $p = 2$, we have the following decomposition

$$\mathrm{su}(2) = \sigma_x \oplus \sigma_y \oplus \sigma_z$$

in terms of vector space sum.

4.3.5. *MUBs and the special linear group*

According to the orthogonal decomposition proposition, in the case where $d = p$ is a prime number (even or odd), the set $\{X^a Z^b \mid a, b \in \mathbb{Z}_p\} \setminus \{X^0 Z^0\}$ of cardinality $p^2 - 1$ can be partitioned into $p+1$ subsets, each containing $p-1$ commuting matrices.

As an example, let us consider the case $d = 5$. For this case, we are left with the six following sets of four commuting matrices

$$\mathcal{V}_0 = \{01, 02, 03, 04\}, \quad \mathcal{V}_1 = \{10, 20, 30, 40\}$$

$$\mathcal{V}_2 = \{11, 22, 33, 44\}, \quad \mathcal{V}_3 = \{12, 24, 31, 43\}$$

$$\mathcal{V}_4 = \{13, 21, 34, 42\}, \quad \mathcal{V}_5 = \{14, 23, 32, 41\}$$

where ab is used as an abbreviation of $X^a Z^b$.

More generally, for $d = p$, with p prime, the $p+1$ sets of $p-1$ commuting matrices are easily seen to be

$$
\begin{aligned}
\mathcal{V}_0 &= \{X^0 Z^a \mid a = 1, 2, \cdots, p-1\} \\
\mathcal{V}_1 &= \{X^a Z^0 \mid a = 1, 2, \cdots, p-1\} \\
\mathcal{V}_2 &= \{X^a Z^a \mid a = 1, 2, \cdots, p-1\} \\
\mathcal{V}_3 &= \{X^a Z^{2a} \mid a = 1, 2, \cdots, p-1\} \\
&\vdots \\
\mathcal{V}_{p-1} &= \{X^a Z^{(p-2)a} \mid a = 1, 2, \cdots, p-1\} \\
\mathcal{V}_p &= \{X^a Z^{(p-1)a} \mid a = 1, 2, \cdots, p-1\}
\end{aligned}
$$

Each of the $p+1$ sets $\mathcal{V}_0, \mathcal{V}_1, \cdots, \mathcal{V}_p$ can be put in a one-to-one correspondence with one basis of the complete set of $p+1$ MUBs. In fact, $\mathcal{V}_0$ is associated with the computational basis, whereas $\mathcal{V}_1, \mathcal{V}_2, \cdots, \mathcal{V}_p$ are associated with the p remaining MUBs in view of

$$
V_a \in \mathcal{V}_{a+1} = \{X^b Z^{ab} \mid b = 1, 2, \cdots, p-1\}, \quad a = 0, 1, \cdots, p-1
$$

More precisely, we have

$$
Z \in \mathcal{V}_0, \quad X \in \mathcal{V}_1, \quad XZ \in \mathcal{V}_2, \quad \cdots, \quad XZ^{p-1} \in \mathcal{V}_p.
$$

The eigenvectors of the $p+1$ unitary operators

$$
Z, \quad X, \quad XZ, \quad \cdots, \quad XZ^{p-1}
$$

generate $p+1$ MUBs (one basis is associated with each of the $p+1$ operators).

4.3.6. *MUBs for d power of a prime*

We may ask what becomes the proposition in section 4.3.1 when the prime number p is replaced by an arbitrary (not prime) number d. In this case, the master formula, with p replaced by d, does not provide a complete set of $d+1$

MUBs. However, it is easy to verify that the bases B_0, B_1 and B_d are three MUBs in $\mathbb{C}^d$, in agreement with the well-known result according to which the number of MUBs in $\mathbb{C}^d$, with d arbitrary, is greater than or equal to 3.

The master formula for $\mathbb{C}^p$ given in 4.3.1 can be used to derive a complete set of $p^m + 1$ MUBs in $\mathbb{C}^{p^m}$ (p prime and $m \geq 2$) by tensor products of order m of vectors in $\mathbb{C}^p$. The general case is very much involved. Hence, we shall limit ourselves to the case $d = 2^2$.

The case $d = 4$ corresponds to the spin angular momentum $j = \frac{3}{2}$. The four bases B_a for $a = 0, 1, 2, 3$ consisting of the vectors $|a\alpha\rangle$ calculated for $d = 4$ from section 4.2.3 and the computational basis B_4 do not constitute a complete set of $d + 1 = 5$ MUBs. Nevertheless, it is possible to find $d + 1 = 5$ MUBs because $d = 2^2$ is the power of a prime number. Indeed, another way to deal with the search for MUBs in $\mathbb{C}^4$ is to consider two systems of qubits associated with the spin angular momenta $j_1 = \frac{1}{2} \Leftrightarrow d_1 = p = 2$ and $j_2 = \frac{1}{2} \Leftrightarrow d_2 = p = 2$. Then, bases of $\mathbb{C}^4$ can be constructed from tensor products $|a\alpha\rangle \otimes |b\beta\rangle$ which are eigenvectors of the operator $v_a \otimes v_b$, where v_a corresponds to the first system of qubits and v_b to the second one. Obviously, the set

$$B_{ab} = \{|a\alpha\rangle \otimes |b\beta\rangle \mid \alpha, \beta = 0, 1\}$$

is an orthonormal basis of $\mathbb{C}^4$. Four of the five MUBs for $d = 2^2 = 4$ can be constructed from the various bases B_{ab}. It is evident that B_{00} and B_{11} are two unbiased bases since the modulus of the inner product of $|1\alpha'\rangle \otimes |1\beta'\rangle$ by $|0\alpha\rangle \otimes |0\beta\rangle$ is

$$|\langle 0\alpha|1\alpha'\rangle\langle 0\beta|1\beta'\rangle| = \frac{1}{\sqrt{4}} = \frac{1}{\sqrt{d}}$$

A similar result holds for the two bases B_{01} and B_{10}. However, the four bases B_{00}, B_{11}, B_{01} and B_{10} are not mutually unbiased. A possible way to overcome this no-go result is to keep the bases B_{00} and B_{11} intact and to re-organize the vectors inside the bases B_{01} and B_{10} in order to obtain four MUBs. We are thus left with the four bases

$$W_{00} \equiv B_{00}, \quad W_{11} \equiv B_{11}, \quad W_{01}, \quad W_{10}$$

which, together with the computational basis B_4, give five MUBs. In detail, we have

$$W_{00} = \{|0\alpha\rangle \otimes |0\beta\rangle \mid \alpha, \beta = 0, 1\}$$

$$W_{11} = \{|1\alpha\rangle \otimes |1\beta\rangle \mid \alpha, \beta = 0, 1\}$$

$$W_{01} = \{\lambda|0\alpha\rangle \otimes |1\beta\rangle + \mu|0\alpha \oplus 1\rangle \otimes |1\beta \oplus 1\rangle \mid \alpha, \beta = 0, 1\}$$

$$W_{10} = \{\lambda|1\alpha\rangle \otimes |0\beta\rangle + \mu|1\alpha \oplus 1\rangle \otimes |0\beta \oplus 1\rangle \mid \alpha, \beta = 0, 1\}$$

where the addition $\oplus$ should be understood modulo 4; furthermore

$$\lambda = \frac{1 - \mathrm{i}}{2}, \quad \mu = \frac{1 + \mathrm{i}}{2}$$

and the vectors of type $|a\alpha\rangle$ are given by the master formula in section 4.3.1. As a result, only two formulas are necessary to obtain the $d^2 = 16$ vectors $|ab; \alpha\beta\rangle$ for the bases W_{ab}, namely

$$W_{00}, W_{11} : |aa; \alpha\beta\rangle = |a\alpha\rangle \otimes |a\beta\rangle$$

$$W_{01}, W_{10} : |aa \oplus 1; \alpha\beta\rangle = \lambda|a\alpha\rangle \otimes |a \oplus 1\beta\rangle + \mu|a\alpha \oplus 1\rangle \otimes |a \oplus 1\beta \oplus 1\rangle,$$

for all a, α, β in $\mathbb{F}_2$. A simple development of W_{00}, W_{11}, W_{01} and W_{10} gives the following expressions.

The W_{00} basis:

$$|00; 00\rangle = \frac{1}{2}(|0\rangle \otimes |0\rangle + |0\rangle \otimes |1\rangle + |1\rangle \otimes |0\rangle + |1\rangle \otimes |1\rangle)$$

$$|00; 01\rangle = \frac{1}{2}(|0\rangle \otimes |0\rangle - |0\rangle \otimes |1\rangle + |1\rangle \otimes |0\rangle - |1\rangle \otimes |1\rangle)$$

$$|00; 10\rangle = \frac{1}{2}(|0\rangle \otimes |0\rangle + |0\rangle \otimes |1\rangle - |1\rangle \otimes |0\rangle - |1\rangle \otimes |1\rangle)$$

$$|00; 11\rangle = \frac{1}{2}(|0\rangle \otimes |0\rangle - |0\rangle \otimes |1\rangle - |1\rangle \otimes |0\rangle + |1\rangle \otimes |1\rangle)$$

or in column vectors

$$\frac{1}{2}\begin{pmatrix}1\\1\\1\\1\end{pmatrix}, \quad \frac{1}{2}\begin{pmatrix}1\\-1\\1\\-1\end{pmatrix}, \quad \frac{1}{2}\begin{pmatrix}1\\1\\-1\\-1\end{pmatrix}, \quad \frac{1}{2}\begin{pmatrix}1\\-1\\-1\\1\end{pmatrix}$$

The W_{11} basis:

$$|11;00\rangle = \frac{1}{2}(|0\rangle \otimes |0\rangle + \mathrm{i}|0\rangle \otimes |1\rangle + \mathrm{i}|1\rangle \otimes |0\rangle - |1\rangle \otimes |1\rangle)$$

$$|11;01\rangle = \frac{1}{2}(|0\rangle \otimes |0\rangle - \mathrm{i}|0\rangle \otimes |1\rangle + \mathrm{i}|1\rangle \otimes |0\rangle + |1\rangle \otimes |1\rangle)$$

$$|11;10\rangle = \frac{1}{2}(|0\rangle \otimes |0\rangle + \mathrm{i}|0\rangle \otimes |1\rangle - \mathrm{i}|1\rangle \otimes |0\rangle + |1\rangle \otimes |1\rangle)$$

$$|11;11\rangle = \frac{1}{2}(|0\rangle \otimes |0\rangle - \mathrm{i}|0\rangle \otimes |1\rangle - \mathrm{i}|1\rangle \otimes |0\rangle - |1\rangle \otimes |1\rangle)$$

or in column vectors

$$\frac{1}{2}\begin{pmatrix}1\\\mathrm{i}\\\mathrm{i}\\-1\end{pmatrix}, \quad \frac{1}{2}\begin{pmatrix}1\\-\mathrm{i}\\\mathrm{i}\\1\end{pmatrix}, \quad \frac{1}{2}\begin{pmatrix}1\\\mathrm{i}\\-\mathrm{i}\\1\end{pmatrix}, \quad \frac{1}{2}\begin{pmatrix}1\\-\mathrm{i}\\-\mathrm{i}\\-1\end{pmatrix}$$

The W_{01} basis:

$$|01;00\rangle = \frac{1}{2}(|0\rangle \otimes |0\rangle + |0\rangle \otimes |1\rangle - \mathrm{i}|1\rangle \otimes |0\rangle + \mathrm{i}|1\rangle \otimes |1\rangle)$$

$$|01;11\rangle = \frac{1}{2}(|0\rangle \otimes |0\rangle - |0\rangle \otimes |1\rangle + \mathrm{i}|1\rangle \otimes |0\rangle + \mathrm{i}|1\rangle \otimes |1\rangle)$$

$$|01;01\rangle = \frac{1}{2}(|0\rangle \otimes |0\rangle - |0\rangle \otimes |1\rangle - \mathrm{i}|1\rangle \otimes |0\rangle - \mathrm{i}|1\rangle \otimes |1\rangle)$$

$$|01;10\rangle = \frac{1}{2}(|0\rangle \otimes |0\rangle + |0\rangle \otimes |1\rangle + \mathrm{i}|1\rangle \otimes |0\rangle - \mathrm{i}|1\rangle \otimes |1\rangle)$$

or in column vectors

$$\frac{1}{2}\begin{pmatrix}1\\1\\-\mathrm{i}\\\mathrm{i}\end{pmatrix},\quad \frac{1}{2}\begin{pmatrix}1\\-1\\\mathrm{i}\\\mathrm{i}\end{pmatrix},\quad \frac{1}{2}\begin{pmatrix}1\\-1\\-\mathrm{i}\\-\mathrm{i}\end{pmatrix},\quad \frac{1}{2}\begin{pmatrix}1\\1\\\mathrm{i}\\-\mathrm{i}\end{pmatrix}$$

The W_{10} basis:

$$|10;00\rangle = \frac{1}{2}(|0\rangle \otimes |0\rangle - \mathrm{i}|0\rangle \otimes |1\rangle + |1\rangle \otimes |0\rangle + \mathrm{i}|1\rangle \otimes |1\rangle)$$

$$|10;11\rangle = \frac{1}{2}(|0\rangle \otimes |0\rangle + \mathrm{i}|0\rangle \otimes |1\rangle - |1\rangle \otimes |0\rangle + \mathrm{i}|1\rangle \otimes |1\rangle)$$

$$|10;01\rangle = \frac{1}{2}(|0\rangle \otimes |0\rangle + \mathrm{i}|0\rangle \otimes |1\rangle + |1\rangle \otimes |0\rangle - \mathrm{i}|1\rangle \otimes |1\rangle)$$

$$|10;10\rangle = \frac{1}{2}(|0\rangle \otimes |0\rangle - \mathrm{i}|0\rangle \otimes |1\rangle - |1\rangle \otimes |0\rangle - \mathrm{i}|1\rangle \otimes |1\rangle)$$

or in column vectors

$$\frac{1}{2}\begin{pmatrix}1\\-\mathrm{i}\\1\\\mathrm{i}\end{pmatrix},\quad \frac{1}{2}\begin{pmatrix}1\\\mathrm{i}\\-1\\\mathrm{i}\end{pmatrix},\quad \frac{1}{2}\begin{pmatrix}1\\\mathrm{i}\\1\\-\mathrm{i}\end{pmatrix},\quad \frac{1}{2}\begin{pmatrix}1\\-\mathrm{i}\\-1\\-\mathrm{i}\end{pmatrix}$$

The computational basis:

$$|0\rangle \otimes |0\rangle, \quad |0\rangle \otimes |1\rangle, \quad |1\rangle \otimes |0\rangle, \quad |1\rangle \otimes |1\rangle$$

or in column vectors

$$\begin{pmatrix}1\\0\\0\\0\end{pmatrix},\quad \begin{pmatrix}0\\1\\0\\0\end{pmatrix},\quad \begin{pmatrix}0\\0\\1\\0\end{pmatrix},\quad \begin{pmatrix}0\\0\\0\\1\end{pmatrix}$$

It is to be noted that the vectors of the bases W_{00} and W_{11} are not intricated (i.e. each vector is the tensor product of two vectors), whereas the vectors of the bases W_{01} and W_{10} are intricated (i.e. each vector is not the tensor product of two vectors). To be more precise, the degree of intrication of the state vectors for the bases W_{00}, W_{11}, W_{01} and W_{10} can be determined in the following way. In arbitrary dimension d, let

$$|\Phi\rangle = \sum_{k=0}^{d-1} \sum_{l=0}^{d-1} a_{kl} |k\rangle \otimes |l\rangle$$

be a state vector for a system consisting of two sub-systems of qudits. Then, it can be shown that the determinant of the $d \times d$ matrix $A = (a_{kl})$ satisfies

$$0 \leq |\det A| \leq \frac{1}{\sqrt{d^d}}$$

The case $\det A = 0$ corresponds to the absence of global intrication, whereas the case

$$|\det A| = \frac{1}{\sqrt{d^d}}$$

corresponds to a maximal intrication. As an illustration, for $d = 4$, we find again that all the state vectors for W_{00} and W_{11} are not intricated and that all the state vectors for W_{01} and W_{10} are maximally intricated.

The generalization of the formulas given above for two systems of qubits can be obtained in more complicated situations (two systems of qupits, three systems of qubits, etc.). The generalization of the bases W_{00} and W_{11} is immediate. The generalization of W_{01} and W_{10} can be achieved by taking linear combinations of vectors such that each linear combination is made of vectors corresponding to the same eigenvalue of the relevant tensor product of operators of type v_a.

4.4. Galois field approach to mutually unbiased bases

The existence of a complete set of $p^m + 1$ MUBS in $\mathbb{C}^{p^m}$ (p prime and m positive integer) is an indication of a possible utility of Galois fields and Galois rings for the construction of MUBs in $\mathbb{C}^{p^m}$ (p prime, $m \geq 2$). Indeed, the passage from the case $d = p$ to the case $d = p^m$ (p prime, $m \geq 2$) can be achieved by considering the Galois field $\mathbb{GF}(p^m)$ for p odd prime and the Galois ring $\mathbb{GR}(2^2, m)$ for $p = 2$. In this section, we shall deal with the construction of a complete set of $p^m + 1$ MUBs in $\mathbb{C}^{p^m}$, corresponding to the case of m qupits, via the use of the Galois field $\mathbb{GF}(p^m)$ for p odd prime and m greater than 1.

4.4.1. *Weyl pair for* $\mathbb{GF}(p^m)$

4.4.1.1. *The computational basis*

We first have to define the computational basis B_{p^m} in the framework of $\mathbb{GF}(p^m)$, p odd prime and $m \geq 2$. The vectors of the basis B_{p^m} of the Hilbert space $\mathbb{C}^{p^m}$ can be labeled by the elements x of the Galois field $\mathbb{GF}(p^m)$. This can be done in two ways according to which the elements x are taken in the monomial form ($x = 0, \ \alpha^\ell$ with $\ell = 1, 2, \cdots, p^m - 1$) or in the polynomial form ($x = [x_0 x_1 \cdots x_{m-1}]$ with $x_0, x_1, \cdots, x_{m-1} \in \mathbb{F}_p$). In both cases, we have

$$B_{p^m} = \{|0\rangle \text{ or } \phi_0, \quad |1\rangle \text{ or } \phi_1, \quad \cdots, \quad |p^m - 1\rangle \text{ or } \phi_{p^m-1}\}$$

in terms of vectors or column vectors. More precisely, this can be achieved as follows.

– In the monomial form, we define the vectors of B_{p^m} via the correspondences

$$x = 0 \mapsto |0\rangle \text{ or } \phi_0, \quad x = \alpha^\ell \mapsto |\ell\rangle \text{ or } \phi_\ell \text{ with } \ell = 1, 2, \cdots, p^m - 1$$

where α is a primitive element of $\mathbb{GF}(p^m)$.

– In the polynomial form, we can range the vectors of B_{p^m} in the order $0, 1, \cdots, p^m - 1$ by adopting the lexicographical order for the elements $[x_0 x_1 \cdots x_{m-1}]$.

These notations are reminiscent of those used for the computational basis

$$B_p = \{|0\rangle \text{ or } \phi_0, \quad |1\rangle \text{ or } \phi_1, \quad \cdots, \quad |p - 1\rangle \text{ or } \phi_{p-1}\}$$

corresponding to the limit case $m = 1$.

4.4.1.2. *Shift and phase operators for* $\mathbb{GF}(p^m)$

The notion of Weyl pair can be extended to any Galois field $\mathbb{GF}(p^m)$ with p (even or odd) prime and $m \geq 2$. Let x and y be two elements of $\mathbb{GF}(p^m)$ and ϕ_y be the basis column vector of B_{p^m} associated with y. For fixed x, we define the matrices $\hat{X}_x$ (shift operators) and $\hat{Z}_x$ (phase operators) via the actions

$$\hat{X}_x \phi_y = \phi_{y-x}, \quad \hat{Z}_x \phi_y = \chi(xy)\phi_y = \mathrm{e}^{\mathrm{i}\frac{2\pi}{p}\mathrm{Tr}(xy)}\phi_y$$

where y is arbitrary. We easily verify the properties

$$\hat{X}_{x+y} = \hat{X}_x \hat{X}_y = \hat{X}_y \hat{X}_x, \quad \hat{Z}_{x+y} = \hat{Z}_x \hat{Z}_y = \hat{Z}_y \hat{Z}_x$$

and

$$\hat{X}_x \hat{Z}_y - \chi(xy)\hat{Z}_y \hat{X}_x = O_{p^m}, \quad \chi(xy) = \mathrm{e}^{\mathrm{i}\frac{2\pi}{p}\mathrm{Tr}(xy)}$$

In the limit case $m = 1$ (i.e. for the base field $\mathbb{F}_p$), the matrices

$$X = \hat{X}_1, \quad Z = \hat{Z}_1$$

corresponding to $x = y = 1$ satisfy

$$XZ - \mathrm{e}^{\mathrm{i}\frac{2\pi}{p}} ZX = O_p$$

to be compared with the relations satisfied by the Weyl pair (X, Z) defined in 4.3.4.1.

4.4.2. *Bases in the frame of* $\mathbb{GF}(p^m)$

4.4.2.1. *Passage from* $d = p$ *to* $d = p^m$

We might use the Weyl pair (X_x, Z_y) defined in the framework of $\mathbb{GF}(p^m)$, see section 4.4.1, to determine a complete set of $p^m + 1$ MUBs in $\mathbb{C}^{p^m}$ in a similar way to that used for $m = 1$ with the help of the matrix V_a for a in $\mathbb{F}_p$. However, it is quicker to start from the alternative formula in 4.3.3 giving MUBs in $\mathbb{C}^p$ in order to generate a formula for $\mathbb{C}^{p^m}$ giving back the alternative formula in $\mathbb{C}^p$ in the limit case $m = 1$. In this direction, a possible way to pass from the basis vector

$$\frac{1}{\sqrt{p}} \sum_{x \in \mathbb{F}_p} \mathrm{e}^{\mathrm{i}\frac{2\pi}{p}(ax+\alpha)x} |x\rangle$$

of $\mathbb{C}^p$ to a basis vector of $\mathbb{C}^{p^m}$ is to replace

$$\mathrm{e}^{\mathrm{i}\frac{2\pi}{p}(ax+\alpha)x}, \quad a, \alpha, x \in \mathbb{F}_p$$

with

$$\chi(ax^2 + \alpha x) = \mathrm{e}^{\mathrm{i}\frac{2\pi}{p}\mathrm{Tr}(ax^2+\alpha x)}, \quad a, \alpha, x \in \mathbb{GF}(p^m)$$

where χ is the canonical additive character of $\mathbb{GF}(p^m)$. This yields the two following propositions.

4.4.2.2. *Bases B_a for $a \in \mathbb{GF}(p^m)$*

PROPOSITION 4.6.– For p odd prime and $m \geq 2$, the set

$$B_a = \{|a\alpha\rangle \mid \alpha \in \mathbb{GF}(p^m)\}$$

where

$$|a\alpha\rangle = \frac{1}{\sqrt{p^m}} \sum_{x \in \mathbb{GF}(p^m)} \mathrm{e}^{\mathrm{i}\frac{2\pi}{p}\mathrm{Tr}(ax^2+\alpha x)} |x\rangle, \quad a \in \mathbb{GF}(p^m)$$

constitutes an orthonormal basis of $\mathbb{C}^{p^m}$.

PROOF.– See the proof of the next proposition. □

Note that for $m = 1$

$$\mathrm{Tr}(ax^2 + \alpha x) = ax^2 + \alpha x$$

so that the vector $|a\alpha\rangle$ coincides with the vector $|a\alpha\rangle'$ derived in section 4.3.3. This explains why we chose to extend the alternative formula (see 4.3.3) valid for $\mathbb{C}^p$ to the case $\mathbb{C}^{p^m}$. Indeed, the same kind of extension to the master formula (see 4.3.1) is not possible since $\mathrm{Tr}[\frac{1}{2}n(p-n)a + n\alpha]$ does not make sense.

4.4.3. *MUBs in the frame of* $\mathbb{GF}(p^m)$

PROPOSITION 4.7.– For p odd prime and $m \geq 2$, the p^m bases B_a, a ranging in $\mathbb{GF}(p^m)$, constitute, with the computational basis B_{p^m}, a complete set of $p^m + 1$ MUBs in $\mathbb{C}^{p^m}$.

PROOF.– Let $|a\alpha\rangle$ and $|b\beta\rangle$ be two vectors belonging to the bases B_a and B_b, respectively. We have

$$\langle a\alpha|b\beta\rangle = \frac{1}{p^m} \sum_{x\in\mathbb{GF}(p^m)} \mathrm{e}^{\mathrm{i}\frac{2\pi}{p}\mathrm{Tr}[(b-a)x^2+(\beta-\alpha)x]}, \quad a, b, \alpha, \beta \in \mathbb{GF}(p^m)$$

By using

$$\left| \sum_{x\in\mathbb{GF}(p^m)} \mathrm{e}^{\mathrm{i}\frac{2\pi}{p}\mathrm{Tr}(ux^2+vx)} \right| = \sqrt{p^m}, \quad u \in \mathbb{GF}(p^m)^*, \quad v \in \mathbb{GF}(p^m)$$

(valid for p odd prime), we obtain

$$|\langle a\alpha|b\beta\rangle| = \begin{cases} \delta(\alpha, \beta) \text{ if } b = a \\ \frac{1}{\sqrt{p^m}} \text{ if } b \neq a \end{cases}$$

or in a compact form

$$|\langle a\alpha|b\beta\rangle| = \delta(a, b)\delta(\alpha, \beta) + \frac{1}{\sqrt{p^m}}[1 - \delta(a, b)]$$

which shows that B_a is an orthonormal basis and that the couple (B_a, B_b) with $b \neq a$ is a couple of unbiased bases. Of course, each basis B_a is unbiased to the computational basis B_{p^m}. We thus end up with a total of $p^m + 1$ MUBs as desired. □

The previous result applies in the limit case $m = 1$ for which we recover the $p + 1$ MUBs in $\mathbb{C}^p$.

4.5. Galois ring approach to mutually unbiased bases

In dimension $d = 2^m$, $m \geq 2$, the use of the Galois field $\mathbb{GF}(2^m)$ to construct a complete set of $2^m + 1$ MUBs in $\mathbb{C}^{2^m}$ according to the method used in section 4.4 for $d = p^m$, p odd prime, would lead to a no-win situation because $\gcd(2, 2^m) \neq 1$ (while $\gcd(2, p^m) = 1$ for p odd prime). For $d = 2^m$, which corresponds to the case of m qubits, we can use the Galois ring $\mathbb{GR}(2^2, m)$ (denoted as R_{4^m} too) to construct MUBs in $\mathbb{C}^{2^m}$.

4.5.1. *Bases in the frame of* $\mathbb{GR}(2^2, m)$

We start with the residue class ring

$$\mathbb{GR}(2^2, m) = \mathbb{Z}_{2^2}[\xi]/\langle P_m(\xi)\rangle$$

where $P_m(x)$ is a monic basic irreducible polynomial of degree m (i.e. its restriction $\overline{P_m(x)} = P_m(x)$ modulo 2 is irreducible over $\mathbb{Z}_2$). The 2^m vectors of the computational basis B_{2^m} are labeled by the 2^m elements of the Teichmüller set T_m associated with the ring $\mathbb{Z}_{2^2}[\xi]/\langle P_m(\xi)\rangle$. Thus

$$B_{2^m} = \{|x\rangle \mid x \in T_m\}$$

(the set T_m and the ring $\mathbb{GR}(2^2, m)$ contain 2^m and 4^m elements, respectively).

PROPOSITION 4.8.– For a and α in T_m, let

$$\begin{aligned}|a\alpha\rangle &= \frac{1}{\sqrt{2^m}} \sum_{x \in T_m} \chi[(a + 2\alpha)x]|x\rangle \\ &= \frac{1}{\sqrt{2^m}} \sum_{x \in T_m} \mathrm{e}^{\mathrm{i}\frac{2\pi}{4}\mathrm{Tr}(ax+2\alpha x)}|x\rangle \\ &= \frac{1}{\sqrt{2^m}} \sum_{x \in T_m} \mathrm{i}^{\mathrm{Tr}(ax+2\alpha x)}|x\rangle\end{aligned}$$

where χ is an additive character vector of $\mathbb{GR}(2^2, m)$ and the trace takes its values in $\mathbb{Z}_4$. For fixed a in T_m, the set

$$B_a = \{|a\alpha\rangle \mid \alpha \in T_m\}$$

constitutes an orthonormal basis of $\mathbb{C}^{2^m}$.

PROOF.– See the proof of the next proposition. □

Note that for $m = 1$

$$\text{Tr}(ax + 2\alpha x) = ax + 2\alpha x$$

so that

$$|a\alpha\rangle = \frac{1}{\sqrt{2}} \sum_{x \in \mathbb{F}_2} \text{i}^{ax+2\alpha x} |x\rangle \quad [4.1]$$

to be compared with the vector

$$|a\alpha\rangle = \frac{1}{\sqrt{2}} \sum_{x \in \mathbb{F}_2} \text{e}^{\text{i}\frac{2\pi}{2}[\frac{1}{2}ax(2-x)+\alpha x]} |1 - x\rangle$$

$$= \frac{1}{\sqrt{2}} \sum_{x \in \mathbb{F}_2} \text{i}^{ax(2-x)+2\alpha x} |1 - x\rangle \quad [4.2]$$

given by the master formula in section 4.3.1. In view of the fact that

$$\text{i}^{ax+2\alpha x} = \text{i}^{ax(2-x)+2\alpha x}$$

for $x = 0$ and $x = 1$, the two vectors $|a\alpha\rangle$ in equations [4.1] and [4.2] are the same up to an interchange of the vectors $|0\rangle$ and $|1\rangle$.

4.5.2. *MUBs in the frame of* $\mathbb{GR}(2^2, m)$

PROPOSITION 4.9.– The 2^m bases B_a, with $m \geq 2$ and a ranging in the Teichüller set T_m associated with the Galois ring $\mathbb{GR}(2^2, m)$, constitute, with the computational basis B_{2^m}, a complete set of $2^m + 1$ MUBs in $\mathbb{C}^{2^m}$.

PROOF.– Let $|a\alpha\rangle$ and $|b\beta\rangle$ be two vectors belonging to the bases B_a and B_b, respectively. We have

$$\langle a\alpha|b\beta\rangle = \frac{1}{2^m} \sum_{x \in T_m} \text{e}^{\text{i}\frac{\pi}{2}\text{Tr}[(b-a+2\beta-2\alpha)x]}$$

By using

$$\left|\sum_{x\in T_m} \mathrm{e}^{\mathrm{i}\frac{\pi}{2}\mathrm{Tr}(ux)}\right| = \begin{cases} 0 \text{ if } u \in 2T_m,\ u \neq 0 \\ 2^m \text{ if } u = 0 \\ \sqrt{2^m} \text{ otherwise} \end{cases}$$

we obtain

$$|\langle a\alpha|b\beta\rangle| = \begin{cases} \delta(\alpha,\beta) \text{ if } b = a \\ \frac{1}{\sqrt{2^m}} \text{ if } b \neq a \end{cases}$$

or in a compact form

$$|\langle a\alpha|b\beta\rangle| = \delta(a,b)\delta(\alpha,\beta) + \frac{1}{\sqrt{2^m}}[1-\delta(a,b)]$$

which shows that B_a is an orthonormal basis and that the couple (B_a, B_b) with $b \neq a$ is a couple of unbiased bases. Of course, each basis B_a is unbiased to the computational basis B_{2^m}. We thus end up with a total of $2^m + 1$ MUBs and we are done. □

The previous result applies in the limit case $m = 1$ for which we can recover the $2+1$ MUBs in $\mathbb{C}^2$.

4.5.3. *One- and two-qubit systems*

4.5.3.1. *One-qubit system*

For $m = 1$, the $2^m = 2$ vectors of the computational basis B_2 are labeled with the help of the two elements of the Teichmüller set $T_1 = \mathbb{Z}_2$ of the Galois ring $\mathbb{GR}(2^2, 1) = \mathbb{Z}_{2^2}$. Thus, the basis B_2 is

$$B_2 : |0\rangle = \begin{pmatrix}1\\0\end{pmatrix},\ |1\rangle = \begin{pmatrix}0\\1\end{pmatrix}$$

The vectors $|a\alpha\rangle$ of the basis B_a $(a \in T_1)$ are given by (see 4.5.1)

$$|a\alpha\rangle = \frac{1}{\sqrt{2}}\sum_{x=0}^{1} \mathrm{i}^{(a+2\alpha)x}|x\rangle, \quad \alpha \in T_1 = \{0,1\}$$

This yields the two unbiased bases

$$B_0 : |00\rangle = \frac{1}{\sqrt{2}}(|0\rangle + |1\rangle), \quad |01\rangle = \frac{1}{\sqrt{2}}(|0\rangle - |1\rangle)$$

$$B_1 : |10\rangle = \frac{1}{\sqrt{2}}(|0\rangle + \mathrm{i}|1\rangle), \quad |11\rangle = \frac{1}{\sqrt{2}}(|0\rangle - \mathrm{i}|1\rangle)$$

which, together with the computational basis B_2, form a complete set of $2 + 1 = 3$ MUBs in $\mathbb{C}^2$. Note that the bases B_0 and B_1 are in agreement (up to phase factors and a rearrangement of the vectors inside B_1) with the bases B_0 and B_1 derived in 4.3.2.1.

4.5.3.2. *Two-qubit system*

For $m = 2$, the $2^m = 4$ vectors of the computational basis B_4 are labeled with the help of the four elements of the Teichmüller set $T_2 = \{0,\ \beta^1,\ \beta^2 = 3 + 3\beta,\ \beta^3 = 1\}$ of the Galois ring $\mathbb{GR}(2^2, 2)$ (here, we use β instead of α in order to avoid confusion with the index α in $|a\alpha\rangle$). Thus, the basis B_4 is

$$B_4 : |0\rangle = \begin{pmatrix} 1 \\ 0 \\ 0 \\ 0 \end{pmatrix}, \ |\beta^1 \text{ or } 1\rangle = \begin{pmatrix} 0 \\ 1 \\ 0 \\ 0 \end{pmatrix}, \ |\beta^2 \text{ or } 2\rangle = \begin{pmatrix} 0 \\ 0 \\ 1 \\ 0 \end{pmatrix}, \ |\beta^3 \text{ or } 3\rangle = \begin{pmatrix} 0 \\ 0 \\ 0 \\ 1 \end{pmatrix}$$

The vectors $|a\alpha\rangle$ of the basis B_a ($a = 0,\ \beta^1$ or $1,\ \beta^2$ or $2,\ \beta^3$ or 3) are given by (see 4.5.1)

$$|a\alpha\rangle = \frac{1}{2} \sum_{x \in T_2} \mathrm{i}^{\mathrm{Tr}(ax + 2\alpha x)} |x\rangle$$

$$\alpha \in T_2 = \{0,\ \beta^1,\ \beta^2 = 3 + 3\beta,\ \beta^3 = 1\}$$

with

$$\mathrm{Tr}(ax + 2\alpha x) = ax + 2\alpha x + \phi(ax + 2\alpha x)$$

where ϕ is the generalized Frobenius map $\mathbb{GR}(2^2, 2) \to \mathbb{GR}(2^2, 2)$. The correspondence between the indexes a, α in $|a\alpha\rangle$ and the elements $0, \beta^1, \beta^2, \beta^3$ of T_2 is as follows

$$0 \leftrightarrow a \text{ or } \alpha = 0, \ \beta^1 \leftrightarrow a \text{ or } \alpha = 1$$

$$\beta^2 \leftrightarrow a \text{ or } \alpha = 2, \ \beta^3 \leftrightarrow a \text{ or } \alpha = 3$$

This yields the four unbiased bases

$$B_0 : |00\rangle = \frac{1}{2}\begin{pmatrix}1\\1\\1\\1\end{pmatrix}, |01\rangle = \frac{1}{2}\begin{pmatrix}1\\-1\\1\\-1\end{pmatrix}, |02\rangle = \frac{1}{2}\begin{pmatrix}1\\1\\-1\\-1\end{pmatrix}, |03\rangle = \frac{1}{2}\begin{pmatrix}1\\-1\\-1\\1\end{pmatrix}$$

$$B_1 : |12\rangle = \frac{1}{2}\begin{pmatrix}1\\-\mathrm{i}\\1\\\mathrm{i}\end{pmatrix}, |11\rangle = \frac{1}{2}\begin{pmatrix}1\\\mathrm{i}\\-1\\\mathrm{i}\end{pmatrix}, |13\rangle = \frac{1}{2}\begin{pmatrix}1\\\mathrm{i}\\1\\-\mathrm{i}\end{pmatrix}, |10\rangle = \frac{1}{2}\begin{pmatrix}1\\-\mathrm{i}\\-1\\-\mathrm{i}\end{pmatrix}$$

$$B_2 : |21\rangle = \frac{1}{2}\begin{pmatrix}1\\1\\-\mathrm{i}\\\mathrm{i}\end{pmatrix}, |22\rangle = \frac{1}{2}\begin{pmatrix}1\\-1\\\mathrm{i}\\\mathrm{i}\end{pmatrix}, |20\rangle = \frac{1}{2}\begin{pmatrix}1\\-1\\-\mathrm{i}\\-\mathrm{i}\end{pmatrix}, |23\rangle = \frac{1}{2}\begin{pmatrix}1\\1\\\mathrm{i}\\-\mathrm{i}\end{pmatrix}$$

$$B_3 : |33\rangle = \frac{1}{2}\begin{pmatrix}1\\\mathrm{i}\\\mathrm{i}\\-1\end{pmatrix}, |32\rangle = \frac{1}{2}\begin{pmatrix}1\\-\mathrm{i}\\\mathrm{i}\\1\end{pmatrix}, |31\rangle = \frac{1}{2}\begin{pmatrix}1\\\mathrm{i}\\-\mathrm{i}\\1\end{pmatrix}, |30\rangle = \frac{1}{2}\begin{pmatrix}1\\-\mathrm{i}\\-\mathrm{i}\\-1\end{pmatrix}$$

We thus end up with $4 + 1 = 5$ bases (B_0 to B_4) which form a complete set of MUBs in $\mathbb{C}^4$. Note that the bases B_0, B_1, B_2 and B_3 coincide with the bases W_{00}, W_{10}, W_{01} and W_{11} derived from tensor products, respectively; for the purpose of comparison, the vectors $|a\alpha\rangle$ are listed in the same order for each of the couples (B_0, W_{00}), (B_1, W_{10}), (B_2, W_{01}) and (B_3, W_{11}), see 4.3.6.

5

Appendix on Number Theory and Group Theory

This chapter deals with some basic elements of number theory and group theory of interest for the four preceding chapters. We limit ourselves to a listing of definitions and classical results as well as examples. The theorems and properties are generally given without proof. For more details, see the references in the sections *Mathematical literature: number theory* and *Mathematical literature: group theory* of the bibliography.

5.1. Elements of number theory

5.1.1. *Euler function*

Let $\varphi(n)$ be the number of integers in the set $\{0, 1, \cdots, n-1\}$ which are co-prime to n (n positive integer). The function

$$\begin{aligned} \varphi : \mathbb{N}_1 &\to \mathbb{N}_1 \\ n &\mapsto \varphi(n) \end{aligned}$$

is called the Euler totient function. Note that $1 \leq \varphi(n) \leq n-1$. Table 5.1 gives some values of the function φ.

For p prime and m positive integer, we have

$$\varphi(p^m) = p^m - p^{m-1} = p^m \left(1 - \frac{1}{p}\right) = p^{m-1}(p-1)$$

which gives

$$\varphi(p) = p - 1$$

for $m = 1$, and

$$\varphi(2^m) = 2^{m-1}$$

for $p = 2$. If

$$n = \prod_{i=1}^{r} p_i^{m_i}$$

with p_i prime number and m_i positive integer ($i = 1, 2, \cdots, r$), then

$$\varphi(n) = n \prod_{i=1}^{r} \left(1 - \frac{1}{p_i}\right) = \prod_{i=1}^{r} p_i^{m_i - 1} (p_i - 1)$$

which gives back the formula for $\varphi(p^m)$ when $n = p^m$.

n	1	2	3	4	5	6	7	8	9	10	11	12	13	14	15	16
$\varphi(n)$	1	1	2	2	4	2	6	4	6	4	10	4	12	6	8	8

Table 5.1. *Some values of the Euler totient function* φ

For any positive integer d, we have the property

$$\sum_{n>0,\ n|d} \varphi(n) = d$$

where the summation on n is extended over all positive divisors of d.

5.1.2. *Möbius function*

The Möbius function μ is defined by

$$\begin{aligned} \mu : \mathbb{N}_1 &\to \{-1, 0, 1\} \\ n &\mapsto \mu(n) \end{aligned}$$

where

$$\mu(n) = \begin{cases} 1 \text{ if } n = 1 \\ (-1)^r \text{ if } n \text{ is the product of } r \text{ distinct primes} \\ 0 \text{ if } n \text{ is divisible by the square of a prime} \end{cases}$$

Table 5.2 gives some values of the Möbius function μ.

n	1	2	3	4	5	6	7	8	9	10	11	12	13	14	15	16
$\mu(n)$	1	–1	–1	0	–1	1	–1	0	0	1	–1	0	–1	1	1	0

Table 5.2. *Some values of the Möbius function* μ

For any positive integer d, we have the property

$$\sum_{n>0,\ n|d} \mu(n) = \begin{cases} 1 \text{ if } d = 1 \\ 0 \text{ if } d > 1 \end{cases}$$

where the summation on n is extended over all positive divisors of d. Furthermore, we have the following relation

$$\sum_{n>0,\ n|d} \frac{1}{n}\mu(n) = \frac{1}{d}\varphi(d)$$

between the Euler and Möbius functions.

5.1.3. *Root of unity*

A solution in $\mathbb{C}$ of the equation

$$x^n - 1 = 0, \quad n \in \mathbb{N}_1$$

is called an n-th *root of unity*. There are n n-th roots of unity, namely

$$\omega_k = \mathrm{e}^{\mathrm{i}\frac{2\pi}{n}k}, \quad k = 1, 2, \cdots, n \text{ or } k = 0, 1, \cdots, n-1$$

They are all distinct and it is clear that

$$x^n - 1 = \prod_{k=0}^{n-1} (x - \omega_k)$$

As an example, for $n = 4$, we have

$$\omega_0 = 1, \quad \omega_1 = \mathrm{i}, \quad \omega_2 = -1, \quad \omega_3 = -\mathrm{i}$$

and

$$\begin{aligned}(x - \omega_0)(x - \omega_1)(x - \omega_2)(x - \omega_3)\\ = (x-1)(x-\mathrm{i})(x+1)(x+\mathrm{i}) = x^4 - 1\end{aligned}$$

Let α be an n-th root of unity. Then, α satisfies $x^n = 1$. It may happen that α also satisfies $x^\ell = 1$, where $1 \leq \ell < n$. The smallest value of ℓ such that $x^\ell = 1$, where $1 \leq \ell \leq n$, is called the order of α. By way of illustration, the 4-th roots of unity 1, i, -1 and $-$i are of order 1, 4, 2 and 4, respectively.

A *primitive n-th root of unity* is a root of $x^n - 1 = 0$ that is not a root of $x^\ell - 1 = 0$ with $0 < \ell < n$. Among the n n-th roots of unity ω_k, the roots for which k is a co-prime to n are primitive n-th roots of unity. Therefore, there are $\varphi(n)$ primitive n-th roots of unity. Thus, if $n = p$ is a prime, there are $p-1$ primitive p-th roots of unity, namely

$$\omega_k = \mathrm{e}^{\mathrm{i}\frac{2\pi}{p}k}, \quad k = 1, 2, \cdots, p-1$$

(for arbitrary n, ω_0 is not a primitive n-th root of unity). For example

– the 4-th roots of unity 1 and -1 are not primitive 4-th roots of unity; only i and $-$i are primitive 4-th roots of unity;

– the root i of $x^8 - 1 = 0$ is not a primitive 8-th root of unity, since it satisfies $x^4 - 1 = 0$;

– the root $\mathrm{e}^{\mathrm{i}\frac{2\pi}{n}}$ of $x^n - 1 = 0$ is a primitive n-th root of unity, since it is impossible to find ℓ with $0 < \ell < n$ such that $x^\ell - 1 = 0$;

– the 7-th roots of unity $\omega_k = \mathrm{e}^{\mathrm{i}\frac{2\pi}{7}k}$, $k = 1, 2, \cdots, 6$ are primitive 7-th roots of unity.

In order to establish contact with Galois fields, note that all the roots of

$$x^{p^m - 1} - 1 = 0, \quad p \text{ prime, } m \text{ positive integer}$$

are not primitive $(p^m - 1)$-th roots of unity. As an example, for $p = 3$ and $m = 2$, the elements 1, -1, i and $-$i are not primitive 8-th roots of unity. The sole primitive 8-th roots of unity are $e^{i\frac{\pi}{4}k}$ with $k = 1$, 3, 5 and 7. More generally, the Galois field $\mathbb{GF}(p^m)$ contains a primitive n-th root of unity if and only if n is a divisor of $p^m - 1$. Indeed,

- if $n \not| \, p^m - 1$, then $\mathbb{GF}(p^m)$ does not have primitive n-th roots of unity;
- if $n \mid p^m - 1$, then $\mathbb{GF}(p^m)$ has $\varphi(n)$ primitive n-th roots of unity.

Finally, note that the set $\{e^{i\frac{2\pi}{n}k} \mid k = 0, 1, \cdots, n-1\}$ of the n n-th roots of unity endowed with the multiplication of $\mathbb{C}$ forms a cyclic group isomorphic to C_n. This group has $\varphi(n)$ generators, namely, the $\varphi(n)$ primitive n-th roots of unity.

5.1.4. *Cyclotomic polynomials*

The n-th *cyclotomic polynomial* $\Phi_n(x)$, with $n \in \mathbb{N}_1$, is the monic polynomial, whose roots in $\mathbb{C}$ are the primitive n-th roots of unity. In other words

$$\Phi_n(x) = \prod_{k, \gcd(k,n)=1} (x - e^{i\frac{2\pi}{n}k})$$

where the product on k from 1 to $n-1$ is restricted to the values of k co-prime to n (in other words, the product on k is extended to the units of the ring $\mathbb{Z}_n$). The degree of $\Phi_n(x)$ is $\varphi(n)$. For example

$$\Phi_3(x) = \prod_{k=1}^{2} (x - e^{i\frac{2\pi}{3}k}) = (x - \omega)(x - \omega^2) = 1 + x + x^2$$

where $\omega = e^{i\frac{2\pi}{3}}$. Similarly, for $n = 6$, we get

$$\Phi_6(x) = \prod_{k=1 \text{ and } 5} (x - e^{i\frac{2\pi}{6}k}) = 1 - x + x^2$$

Table 5.3 gives the cyclotomic polynomials $\Phi_n(x)$ with their degree $\varphi(n)$ for $n = 1$ to 16.

n	$\varphi(n)$	$\Phi_n(x)$
1	1	$-1 + x$
2	1	$1 + x$
3	2	$\sum_{n=0 \text{ to } 2} x^n$
4	2	$1 + x^2$
5	4	$\sum_{n=0 \text{ to } 4} x^n$
6	2	$1 - x + x^2$
7	6	$\sum_{n=0 \text{ to } 6} x^n$
8	4	$1 + x^4$
9	6	$1 + x^3 + x^6$
10	4	$1 - x + x^2 - x^3 + x^4$
11	10	$\sum_{n=0 \text{ to } 10} x^n$
12	4	$1 - x^2 + x^4$
13	12	$\sum_{n=0 \text{ to } 12} x^n$
14	6	$1 - x + x^2 - x^3 + x^4 - x^5 + x^6$
15	8	$1 - x + x^3 - x^4 + x^5 - x^7 + x^8$
16	8	$1 + x^8$

Table 5.3. *Cyclotomic polynomials $\Phi_n(x)$ with $\varphi(n)$ for $n = 1$ to 16*

The polynomial $x^n - 1$, with $n \in \mathbb{N}_1$, can be decomposed as

$$x^n - 1 = \prod_{k,\, k|n} \Phi_k(x)$$

where the product on k from 1 to n is restricted to the values of k that divide n. As an example

$$\begin{aligned} x^{12} - 1 &= \prod_{k=1,2,3,4,6,12} \Phi_k(x) \\ &= (-1 + x)(1 + x)(1 + x + x^2)(1 + x^2)(1 - x + x^2) \\ &\quad \times (1 - x^2 + x^4) \end{aligned}$$

5.1.5. *Residue*

The Euclidean division of a (the dividend in $\mathbb{Z}$) by n (the divisor in $\mathbb{N}_1$) yields

$$a = n \times q + r, \quad q \in \mathbb{Z}, \quad 0 \leq r \leq n - 1$$

where r and q are called the *residue* modulo n and the quotient of a by n, respectively. The notation

$$a \equiv r \bmod n$$

is often used in place of $a = n \times q + r$. For example, $25 \equiv 1 \bmod 3$ and $-25 \equiv 2 \bmod 3$ mean $25 = 3 \times 8 + 1$ and $-25 = 3 \times (-9) + 2$, respectively.

Two integers a and b are said to be congruent modulo n if $a \equiv b \bmod n$ or, equivalently, if n divides $a - b$. If the residues modulo n relative to two integers a and b are equal, then a and b are congruent modulo n (we say that a is congruent to b modulo n and reciprocally).

To each of the n possible residues $(0, 1, \cdots, n-1)$ corresponding to the division of a by n, we can associate an equivalence class containing an infinity of elements. The n *residue classes* associated with $0, 1, \cdots, n-1$, namely

$$\begin{aligned}
\mathcal{C}_0 &= \{\cdots, -3n, -2n, -n, 0, n, 2n, 3n, \cdots\} \\
\mathcal{C}_1 &= \{\cdots, 1-3n, 1-2n, 1-n, 1, 1+n, 1+2n, 1+3n, \cdots\} \\
&\vdots \\
\mathcal{C}_{n-1} &= \{\cdots, -1-2n, -1-n, -1, n-1, 2n-1, 3n-1, \cdots\}
\end{aligned}$$

make it possible to partition $\mathbb{Z}$ into n equivalence classes as

$$\mathbb{Z} = \mathcal{C}_0 \cup \mathcal{C}_1 \cup \cdots \cup \mathcal{C}_{n-1} = \bigcup_{k=0}^{n-1} \mathcal{C}_k$$

The set $\{\mathcal{C}_0, \mathcal{C}_1, \cdots, \mathcal{C}_{n-1}\}$ of the n residue classes is denoted as $\mathbb{Z}_n$ or $\mathbb{Z}/n\mathbb{Z}$. Thus

$$\mathbb{Z}_n = \{0, 1, \cdots, n-1\}$$

by identifying the residue class C_r with r $(0 \leq r \leq n-1)$. For instance, for $n = 4$, we have the following partition of $\mathbb{Z}$

$$\mathbb{Z} = C_0 \cup C_1 \cup C_2 \cup C_3$$

with

$$C_0 = \{\cdots, -12, -8, -4, 0, 4, 8, 12, \cdots\}$$
$$C_1 = \{\cdots, -11, -7, -3, 1, 5, 9, 13, \cdots\}$$
$$C_2 = \{\cdots, -10, -6, -2, 2, 6, 10, 14, \cdots\}$$
$$C_3 = \{\cdots, -9, -5, -1, 3, 7, 11, 15, \cdots\}$$

(note that the only even prime number is in C_2 and that the odd prime numbers are in C_1 and C_3).

The notion of residue also applies to the Euclidean division of polynomials (see the residue classes of polynomials occurring in Chapters 2 and 3).

5.1.6. *Quadratic residue*

For fixed a in $\mathbb{F}_p{}^*$, a is a *quadratic residue* if

$$\exists x \in \mathbb{F}_p \mid a \equiv x^2 \bmod p$$

(the integer a is the square of an element x of $\mathbb{F}_p{}^*$ modulo p). In the opposite case, a is called a *quadratic non-residue* (the integer a is not the square of an element x of $\mathbb{F}_p{}^*$ modulo p). This can be described by the Legendre symbol

$$\left(\frac{a}{p}\right) = \begin{cases} +1 \text{ if } a \text{ is a square in } \mathbb{F}_p{}^* \\ -1 \text{ if } a \text{ is not a square in } \mathbb{F}_p{}^* \end{cases}$$

or, in other words,

$$\left(\frac{a}{p}\right) = \begin{cases} +1 \text{ if } \exists x \in \mathbb{F}_p{}^* \mid a = x^2 \\ -1 \text{ if } \forall x \in \mathbb{F}_p{}^* : x^2 \neq a \end{cases}$$

In order to cover the case $a = 0$, we define

$$\left(\frac{a}{p}\right) = 0 \text{ if } a \text{ is the zero element of the field } \mathbb{F}_p$$

We summarize as follows

$$\left(\frac{a}{p}\right) = \begin{cases} 0 \text{ if } a \equiv 0 \bmod p \ (a \text{ is divisible by } p) \\ +1 \text{ if } a \equiv x^2 \bmod p \ (a \text{ is a non-zero quadratic residue}) \\ -1 \text{ if } a \not\equiv x^2 \bmod p \ (a \text{ is a non-zero quadratic non-residue}) \end{cases}$$

where p is a prime. Of course,

$$\left(\frac{a}{p}\right) = \left(\frac{b}{p}\right)$$

if $b \equiv a \bmod p$.

For p odd prime, there are $\frac{p-1}{2}$ non-zero quadratic residues and $\frac{p-1}{2}$ non-zero quadratic non-residues. Thus, in ${\mathbb{F}_p}^*$, half of the elements are quadratic residues and half are quadratic non-residues. Consequently,

$$\sum_{x \in \mathbb{F}_p} \left(\frac{x}{p}\right) = 0$$

for p odd prime.

As a trivial example, in $\mathbb{F}_5$, we have

$$1^2 = 1, \quad 2^2 = 4, \quad 3^2 \equiv 4, \quad 4^2 \equiv 1 \pmod 5$$

Therefore, 1 and 4 are non-zero quadratic residues, whereas 2 and 3 are non-zero quadratic non-residues.

For p odd prime and a integer, we have the Euler formula

$$\left(\frac{a}{p}\right) \equiv a^{\frac{p-1}{2}} \bmod p$$

This yields the Euler criterion: the element a of $\mathbb{F}_p{}^*$ is a quadratic residue or a quadratic non-residue according to whether $a^{\frac{p-1}{2}} \equiv +1 \bmod p$ or $a^{\frac{p-1}{2}} \equiv -1 \bmod p$, respectively. Two particular cases of the Euler formula are of interest. First, for $a = -1$ and p odd prime, we have

$$\left(\frac{-1}{p}\right) = (-1)^{\frac{p-1}{2}} = \begin{cases} +1 \text{ for } p \equiv 1 \bmod 4 \\ -1 \text{ for } p \equiv 3 \bmod 4 \end{cases}$$

Therefore, -1 is a quadratic residue for $p = 1 + 4k$ or a quadratic non-residue for $p = 3 + 4k$, respectively (with k integer such that p is a prime). Second, we have

$$\left(\frac{1}{p}\right) = (+1)^{\frac{p-1}{2}} = 1$$

so that 1 is a quadratic residue for any value of p (odd) prime as is evident from $1 = 1^2$.

A trivial consequence of the Euler formula is

$$\left(\frac{ab}{p}\right) = \left(\frac{a}{p}\right)\left(\frac{b}{p}\right)$$

for all the integers a and b in $\mathbb{F}_p$ with p odd prime. Therefore, the product of two quadratic residues or two quadratic non-residues is a quadratic residue and the product of a quadratic residue by a quadratic non-residue is a quadratic non-residue.

As an illustration of the Euler formula, we obtain

$$\left(\frac{1}{7}\right) = 1, \quad \left(\frac{2}{7}\right) = 8 \equiv 1, \quad \left(\frac{4}{7}\right) = 64 \equiv 1 \pmod 7$$

so that 1, 2 and 4 are quadratic residues; this result can be checked from

$$1 = 1^2 \equiv 6^2, \quad 2 \equiv 3^2 \equiv 4^2, \quad 4 = 2^2 \equiv 5^2 \pmod 7$$

Similarly, we obtain

$$\left(\frac{3}{7}\right) = 27 \equiv -1, \quad \left(\frac{5}{7}\right) = 125 \equiv -1, \quad \left(\frac{6}{7}\right) = 216 \equiv -1 \pmod 7$$

so that 3, 5 and 6 are quadratic non-residues. A third example is

$$\left(\frac{-4}{17}\right) \equiv (-4)^8 \equiv 1 \bmod 17, \quad \left(\frac{-3}{17}\right) \equiv (-3)^8 \equiv -1 \bmod 17$$

which shows that -4 and -3 are quadratic residue and quadratic non-residue modulo 17, respectively.

Note that for $a = 2$ and p odd prime, the Euler formula gives

$$\left(\frac{2}{p}\right) = (-1)^{\frac{p^2-1}{8}} = \begin{cases} +1 \text{ for } p \equiv \pm 1 \bmod 8 \Rightarrow +1 \text{ for } p \equiv 1 \text{ or } 7 \bmod 8 \\ -1 \text{ for } p \equiv \pm 3 \bmod 8 \Rightarrow -1 \text{ for } p \equiv 3 \text{ or } 5 \bmod 8 \end{cases}$$

from which we recover $\left(\frac{2}{7}\right) = 1$.

For p odd prime and a integer, the equation

$$x^2 - a \equiv 0 \bmod p$$

has

- either 0 solution in $\mathbb{F}_p$
- or 1 solution in $\mathbb{F}_p$ if $a \equiv 0 \bmod p$ (the solution is $x = 0$)
- or 2 solutions in $\mathbb{F}_p$ (if x is a solution, then the other solution is $p - x$).

Thus, the number of solutions to the congruence $x^2 - a \equiv 0 \bmod p$ in $\mathbb{F}_p$ is $1 + \left(\frac{a}{p}\right)$.

For p even prime ($p = 2$) and a integer, the equation $x^2 - a \equiv 0 \bmod 2$ admits one solution in $\mathbb{F}_2$. (This is in accord with the result that a polynomial equation of degree $n \in \mathbb{N}_1$, with coefficients in a field $\mathbb{K}$, admits at most n distinct roots in $\mathbb{K}$. This result is not valid if $\mathbb{K}$ is replaced by an arbitrary ring.)

For p and $q \neq p$ odd primes, we have

$$\left(\frac{q}{p}\right) = (-1)^{\frac{p-1}{2}\frac{q-1}{2}} \left(\frac{p}{q}\right) \Leftrightarrow \left(\frac{q}{p}\right)\left(\frac{p}{q}\right) = (-1)^{\frac{p-1}{2}\frac{q-1}{2}}$$

which is known as the Legendre-Gauss law of quadratic reciprocity. Therefore, if p or $q \equiv 1 \bmod 4$, then

$$\left(\frac{q}{p}\right) = \left(\frac{p}{q}\right)$$

while if p and $q \equiv 3 \bmod 4$, then

$$\left(\frac{q}{p}\right) = -\left(\frac{p}{q}\right)$$

5.1.7. *Gauss sums*

5.1.7.1. *Gauss sums over $\mathbb{Z}_d$*

The quadratic sum

$$G(d) = \sum_{n \in \mathbb{Z}_d} \mathrm{e}^{\mathrm{i}\frac{2\pi}{d}n^2}, \quad d \in \mathbb{N}_1$$

is called the Gauss sum. Gauss proved that

$$\begin{aligned} G(d) = (1+\mathrm{i})\sqrt{d} \quad & \text{if} \quad d \equiv 4 \bmod 4 \\ \mathrm{i}\sqrt{d} \quad & \text{if} \quad d \equiv 3 \bmod 4 \\ 0 \quad & \text{if} \quad d \equiv 2 \bmod 4 \\ \sqrt{d} \quad & \text{if} \quad d \equiv 1 \bmod 4 \end{aligned}$$

or in a compact form

$$\sum_{n \in \mathbb{Z}_d} \mathrm{e}^{\mathrm{i}\frac{2\pi}{d}n^2} = \frac{1+\mathrm{i}^{-d}}{1+\mathrm{i}^{-1}}\sqrt{d}, \quad d \in \mathbb{N}_1$$

Note that the summation $\sum_{n \in \mathbb{Z}_d}$ in $G(d)$ can be written as $\sum_{n=0}^{d-1}$ or $\sum_{n=1}^{d}$.

For p odd prime and a in $\mathbb{F}_p{}^*$ (a and p are co-prime), the Gauss sum $G(p)$ can be expressed in terms of Legendre symbols as

$$G(p) = \sum_{x \in \mathbb{F}_p} \mathrm{e}^{\mathrm{i}\frac{2\pi}{p}x^2} = \sum_{n=0}^{p-1} \left(\frac{n}{p}\right) \mathrm{e}^{\mathrm{i}\frac{2\pi}{p}n} = \sum_{n=1}^{p-1} \left(\frac{n}{p}\right) \mathrm{e}^{\mathrm{i}\frac{2\pi}{p}n}$$

The proof is based on the fact that $1+(\frac{n}{p})$ is the number of solutions to $x^2-n \equiv 0$ modulo p in $\mathbb{F}_p$ (the summation $\sum_{n=0}^{p-1}$ can be replaced by $\sum_{n=1}^{p-1}$, since $(\frac{0}{p}) = 0$).

As an extension of $G(d)$, let us consider the generalized Gauss sum

$$G_{a,b}(d) = \sum_{x \in \mathbb{Z}_d} \mathrm{e}^{\mathrm{i}\frac{2\pi}{d}(ax^2+bx)}, \quad d \in \mathbb{N}_1, \quad a, b \in \mathbb{Z}, \quad (a, d) = 1$$

The following particular cases are of interest.

– For $a = 1$ and $b = 0$, we have

$$G_{1,0}(d) = G(d)$$

– For $d = p$ odd prime, $(a, p) = 1$ and $b = 0$, we have

$$G_{a,0}(p) = \sum_{x \in \mathbb{F}_p} \mathrm{e}^{\mathrm{i}\frac{2\pi}{p}ax^2} = \sum_{n=0}^{p-1} \left(\frac{n}{p}\right) \mathrm{e}^{\mathrm{i}\frac{2\pi}{p}an} = \sum_{n=1}^{p-1} \left(\frac{n}{p}\right) \mathrm{e}^{\mathrm{i}\frac{2\pi}{p}an}$$

Furthermore,

$$G_{a,0}(p) = \left(\frac{a}{p}\right) G(p)$$

In detailed form

$$\sum_{x \in \mathbb{F}_p} \mathrm{e}^{\mathrm{i}\frac{2\pi}{p}ax^2} = \left(\frac{a}{p}\right) \sum_{x \in \mathbb{F}_p} \mathrm{e}^{\mathrm{i}\frac{2\pi}{p}x^2} = \left(\frac{a}{p}\right) \sum_{n=1}^{p-1} \left(\frac{n}{p}\right) \mathrm{e}^{\mathrm{i}\frac{2\pi}{p}n}$$

Therefore

$$\sum_{x \in \mathbb{F}_p} \mathrm{e}^{\mathrm{i}\frac{2\pi}{p}ax^2} = \begin{cases} \left(\frac{a}{p}\right) \sqrt{p} \text{ for } p \equiv 1 \bmod 4 \\ \left(\frac{a}{p}\right) \mathrm{i}\sqrt{p} \text{ for } p \equiv 3 \bmod 4 \end{cases}$$

so that

$$\left| \sum_{x \in \mathbb{F}_p} \mathrm{e}^{\mathrm{i}\frac{2\pi}{p}ax^2} \right| = \sqrt{p}$$

or equivalently

$$|G_{a,0}(p)| = |G(p)| = \sqrt{p}$$

for p odd prime and $(a, p) = 1$ ($\Rightarrow$ $a = 1, 2, \cdots, p-1$ modulo p).

– More generally, we have

$$\left| \sum_{x \in \mathbb{F}_p} e^{i\frac{2\pi}{p}(ax^2+bx)} \right| = \sqrt{p}$$

for p odd prime, $(a, p) = 1$ and $b \in \mathbb{Z}$. (To pass from $|G_{a,b}(p)|$ to $|G(p)|$, replace n by $n + t$ in $G_{a,b}(p)$, where t is a residue modulo p, and take t such as $2at + b = 0$ modulo p. This yields $|G_{a,b}(p)| = |G(p)|$.)

5.1.7.2. *Reciprocity formula for generalized Gauss sums*

The reciprocity formula is

$$\sum_{n=0}^{|c|-1} e^{i\frac{\pi}{c}(an^2+bn)} = \sqrt{\left|\frac{c}{a}\right|}\, e^{i\frac{\pi}{4ac}(|ac|-b^2)} \sum_{n=0}^{|a|-1} e^{-i\frac{\pi}{a}(cn^2+bn)}$$

where a, b and c are the integers such that

$$ac \neq 0, \quad ac + b \text{ even}$$

By putting

$$S(a, b, c) = \sum_{n=0}^{|c|-1} e^{i\frac{\pi}{c}(an^2+bn)}$$

the reciprocity formula reads

$$S(a, b, c) = \sqrt{\left|\frac{c}{a}\right|}\, e^{i\frac{\pi}{4ac}(|ac|-b^2)} S(c, b, -a)$$

Note the symmetry properties

$$S(a, b, c) = S(-a, -b, -c) = S(a, -b, c)$$

5.1.7.3. *Particular Gauss sums*

The reciprocity formula

$$S(a,b,c) = \sqrt{\left|\frac{c}{a}\right|}\, \mathrm{e}^{\mathrm{i}\frac{\pi}{4ac}(|ac|-b^2)} S(-c,-b,a)$$

(with a, b and c integers such that $ac \neq 0$ and $ac+b$ even) can be exploited to obtain the value of some Gauss sums.

– For $a = 2$, $b = 0$ and $c = d$ (d positive integer), we have

$$\sum_{n=0}^{d-1} \mathrm{e}^{\mathrm{i}\frac{2\pi}{d}n^2} = \frac{1}{2}(1+\mathrm{i})\left(1+\frac{1}{\mathrm{i}^d}\right)\sqrt{d}$$

from which we recover

$$\sum_{n\in\mathbb{Z}_d} \mathrm{e}^{\mathrm{i}\frac{2\pi}{d}n^2} = \frac{1+\mathrm{i}^{-d}}{1+\mathrm{i}^{-1}}\sqrt{d}$$

– For $a = 1$, $b = 0$ and $c = d$ (d even positive integer), we obtain

$$\sum_{n=0}^{d-1} \mathrm{e}^{\mathrm{i}\frac{\pi}{d}n^2} = \mathrm{e}^{\mathrm{i}\frac{\pi}{4}}\sqrt{d} \Rightarrow \left|\sum_{n=0}^{d-1} \mathrm{e}^{\mathrm{i}\frac{\pi}{d}n^2}\right| = \sqrt{d}$$

– For $a = b = 1$ and $c = d$ (d odd positive integer), we obtain

$$\sum_{n=0}^{d-1} \mathrm{e}^{\mathrm{i}\frac{\pi}{d}n(n+1)} = \mathrm{e}^{\mathrm{i}\frac{\pi}{4d}(d-1)}\sqrt{d} \Rightarrow \left|\sum_{n=0}^{d-1} \mathrm{e}^{\mathrm{i}\frac{\pi}{d}n(n+1)}\right| = \sqrt{d}$$

– For $a = b = 2$ and $c = d$ (d positive integer), we derive

$$\sum_{n=0}^{d-1} \mathrm{e}^{\mathrm{i}\frac{2\pi}{d}n(n+1)} = \frac{1}{\sqrt{2}}\mathrm{e}^{\mathrm{i}\frac{\pi}{4d}(d-2)}\left(1-\mathrm{e}^{-\mathrm{i}\frac{\pi}{2}d}\right)\sqrt{d}$$

$$\Rightarrow \left|\sum_{n=0}^{d-1} \mathrm{e}^{\mathrm{i}\frac{2\pi}{d}n(n+1)}\right| = \sqrt{2}\left|\sin\left(\tfrac{\pi}{4}d\right)\right|\sqrt{d}$$

– For $a = 1$ and $b = c = d$ (d positive integer), we find

$$\sum_{n=0}^{d-1} (-1)^n \mathrm{e}^{\mathrm{i}\frac{\pi}{d}n^2} = \mathrm{e}^{\mathrm{i}\frac{\pi}{4}(1-d)}\sqrt{d} \Rightarrow \left|\sum_{n=0}^{d-1} (-1)^n \mathrm{e}^{\mathrm{i}\frac{\pi}{d}n^2}\right| = \sqrt{d}$$

5.2. Elements of group theory

5.2.1. *Axioms of group*

DEFINITION 5.1.– A non-empty set G endowed with one internal composition law, say τ, such that

– the law τ is associative

– there exists a neutral element in G for the law τ and

– each element of G has an inverse with respect to the law τ

is a *group* denoted as (G, τ) or simply G when the knowledge of the nature of the law τ is not relevant. In other words, we have the axioms

1) $\forall R \in G,\ \forall S \in G : \exists T \text{ (unique)} \in G \mid R\tau S = T$

2) $\forall R \in G,\ \forall S \in G,\ \forall T \in G : R\tau(S\tau T) = (R\tau S)\tau T$

3) $\exists E \in G \mid \forall R \in G : R\tau E = E\tau R = R$

4) $\forall R \in G,\ \exists R^{-1} \in G \mid R\tau R^{-1} = R^{-1}\tau R = E$

The element E is called *neutral* or *identity* or *unity element* of G and, for a given element R of G, R^{-1} is called the *inverse element* of R.

It can be shown that the system of axioms 1–4 is redundant: in place of axioms 3 and 4, it is sufficient to assume the existence of a right (respectively, left) neutral element and a right (respectively, left) inverse for each element.

The element E and the inverse R^{-1} of any element R are unique.

For n in $\mathbb{N}_0$, the power R^n of the element R is defined by

$$R^0 = E, \quad R^n = R\tau R\tau \cdots \tau R$$

where R^n contains n factors. For simplicity purposes, very often the (product or combined) element $R\tau S$ is simply written as RS. In many physical and mathematical applications, the law τ can be an addition $+$ (then, often the neutral element E is denoted as 0 and the inverse R^{-1} of R as $-R$), or a multiplication $\times$ (then, the neutral element is often denoted as 1 or I), or the composition of transformations (like symmetries), etc.

If the law τ is commutative, the group (G, τ) is said to be commutative or Abelian. The structure of Abelian group is an essential ingredient (as an additive group with τ standing for $+$) in the structure of ring. In the structure of field, two kinds of groups are used: an additive Abelian group with the law $+$ for τ and a multiplicative (not necessarily commutative) group with the law $\times$ for τ.

The cardinal of G, denoted as $|G|$, can be finite (then, $|G|$ is called the order of G) or infinite. In broad terms, there are finite groups (with a finite number of elements) and infinite groups (with an infinite number of elements). Except for a few examples, we shall not consider infinite discrete and continuous groups, and more specifically Lie groups. In this book, we are mainly concerned with finite groups.

5.2.2. *Direct product of groups*

DEFINITION 5.2.– Let (G, τ) and (G', τ') be two groups. The set

$$G \times G' = \{(R, R') \mid R \in G, \ R' \in G'\}$$

endowed with the law σ defined by

$$(R, R')\sigma(S, S') = (R\tau S, R'\tau' S')$$

is a group denoted as $(G \times G', \sigma)$ or simply $G \times G'$ and called the *direct product* of (G, τ) by (G', τ'). If (G, τ) and (G', τ') are finite groups (of order $|G|$ and $|G'|$, respectively), then $(G \times G', \sigma)$ is of order $|G| \, |G'|$.

5.2.3. *Homomorphism, isomorphism and automorphism of groups*

DEFINITION 5.3.– Let (G, τ) and (G', τ') be two groups. An application

$$\begin{aligned} f : (G, \tau) &\rightarrow (G', \tau') \\ R &\mapsto f(R) \end{aligned}$$

such that

$$\forall R \in G,\ \forall S \in G : f(R \tau S) = f(R) \tau' f(S)$$

is called an homomorphism of (G, τ) into (G', τ'). If G and G' have the same cardinal and if f is one-to-one, the application f is called an isomorphism of (G, τ) onto (G', τ') and the groups (G, τ) and (G', τ') are said to be isomorphic. An isomorphism of a group onto itself is called an automorphism.

It is evident that $f(E)$ is the neutral element of G' and $f(R^{-1}) = f(R)^{-1}$.

5.2.4. *Conjugate classes*

DEFINITION 5.4.– Let R and S be two elements of a group (G, τ). If there exists an element T of G such that

$$S = T \tau R \tau T^{-1}$$

the elements R and S are said to be conjugate, and the set

$$\mathcal{C}_R = \{T \tau R \tau T^{-1} \mid T \in G\}$$

is called the *conjugate class* associated with R. The relation $S = T \tau R \tau T^{-1}$ defines an equivalence relation on G. Therefore, the set G can be partitioned into conjugate classes.

It is clear that each conjugate class of an Abelian group contains one and only one element. Therefore, the number of conjugate classes of an Abelian finite group is equal to the order of the group.

From now on, when no confusion is possible, we do not mention the group law τ and we denote the elements of a group with lower case letters $a, b, c, \cdots$ (the letter e being reserved for the neutral element).

5.2.5. *Sub-group*

DEFINITION 5.5.– Let (G, τ) be a group and H a non-empty subset of G. If the set H endowed with the law τ is a group, then (H, τ) is said to be a *sub-group* of (G, τ). The notation $(H, \tau) \subset (G, \tau)$, or simply $H \subset G$, is used to indicate that (H, τ) is a sub-group of (G, τ).

The neutral element of (G, τ) is also the neutral element of its sub-group (H, τ).

The groups (G, τ) and $(\{e\}, \tau)$, where e is the neutral element of G, are trivial or *improper sub-groups* of (G, τ). The other sub-groups of (G, τ), if any, are called non-trivial or *proper sub-groups* of (G, τ).

The two following results illustrate the notion of sub-group.

– Let us consider the direct product $(G \times G', \sigma)$ of the groups (G, τ) and (G', τ'). The subset $\{(x, e') \mid x \in G\}$ of $G \times G'$, endowed with the product law σ, is a sub-group of the group $(G \times G', \sigma)$. This sub-group is isomorphic to G. A similar result holds for the subset $\{(e, x') \mid x' \in G'\}$ of $G \times G'$.

– Let (H, τ) be a sub-group of the group (G, τ) and, for fixed a in G, let

$$aHa^{-1} = \{a\tau x\tau a^{-1} \mid x \in H\}$$

be a subset of G. Then, (aHa^{-1}, τ) is a sub-group of G called the *conjugate sub-group* of (H, τ) in (G, τ) with respect to a.

5.2.6. *Cyclic group*

DEFINITION 5.6.– A finite group for which each element is the power of one of them is said to be cyclic.

In other words, a finite group G is cyclic if there exists one element a of G such that each element of the group is a power of a. The elements of G of order $|G|$ can be written as $a, a^2, \cdots, a^{|G|} = e$ or $a^0 = e, a, \cdots, a^{|G|-1}$. The element a of G is called a generator of G. The cyclic group G generated by the element a is also denoted as $\langle a \rangle$. For any positive integer k, a^k is equally well a generator of $G = \langle a \rangle$ if and only if $\gcd(k, |G|) = 1$. The number of generators of the cyclic finite group G is $\varphi(|G|)$, where φ is the Euler function.

All cyclic finite groups of the same order are isomorphic. It is appropriate to denote C_d the cyclic finite group of order d. Note that the cyclic group C_d, with d an arbitrary positive integer, is isomorphic to the group of the d-th roots of unity, the group law being the multiplication of complex numbers, and to the group $(\mathbb{Z}_d, +)$, the law $+$ being the addition modulo d. When $d = p$ is a prime number, C_p is isomorphic to the additive group $(\mathbb{F}_p, +)$ relative to the field $\mathbb{F}_p$.

Any cyclic group is Abelian. The reverse is evidently false.

Any finite group of prime order is cyclic. The reverse is evidently false.

A proper sub-group, if any, of the cyclic group C_d is a cyclic group (its order is a divisor of d, see Lagrange's theorem in 5.2.8).

The notion of cyclic group exists for infinite group too. All cyclic infinite groups are isomorphic to the group $(\mathbb{Z}, +)$.

5.2.7. *Cosets*

DEFINITION 5.7.– Let G be a group and H a sub-group of G with respect to the law τ and g be an element of G. The set

$$gH = \{g\tau h \mid h \in H\}$$

is called a *left coset* of G with respect to H or, more precisely, the left coset of the element g of G with respect to H. Similarly, the set

$$Hg = \{h\tau g \mid h \in H\}$$

is called a *right coset* of G with respect to H or, more precisely, the right coset of the element g of G with respect to H.

Of course, $eH = He = H$ but in general, for $g \neq e$, we have $gH \neq Hg$ except if G is Abelian. Note that a left (or right) coset of G with respect to H is not a sub-group of G except for the trivial coset H.

Two left (or right) cosets of G with respect to H either have no elements in common or coincide. Any element of G either belongs to H or to a left (or right) coset distinct from H. Therefore, G is a union of disjoint left or right cosets. Thus

$$G = \bigcup_{g \in G} gH = \bigcup_{g \in G} Hg$$

where the union over g is restricted to those values of g yielding distinct cosets.

5.2.8. *Lagrange's theorem*

PROPOSITION 5.1.– The order $|H|$ of a sub-group H of a finite group G is a divisor of the order $|G|$ of G.

PROOF.– If G is finite, then G is the union of a finite number of left or right cosets. For instance,

$$G = H \cup a_2 H \cup a_3 H \cup \cdots \cup a_n H$$

where $H, \; a_2 H, \; a_3 H, \; \cdots, \; a_n H$ are n disjoint left cosets of G with respect to H. In terms of orders, we have

$$|G| = n|H|$$

so that the order of H divides the order of G. □

5.2.9. *Order of a group element*

DEFINITION 5.8.– Let a be an element of a finite group G. The smallest positive integer $n \leq |G|$ such that $a^n = e$ is called the order of a.

If the element a of the group G is of order n, then the set $\{a, a^2, \cdots, a^n\}$, endowed with the law of G, is a cyclic sub-group of G of order n. As a corollary of Lagrange's theorem, the order of any element of a finite group G divides the order $|G|$ of G. Furthermore, for any element a of G, we have $a^{|G|} = e$.

5.2.10. *Quotient group*

DEFINITION 5.9.– Let G be a group and H a sub-group of G such that

$$\forall g \in G : gH = Hg \Leftrightarrow gHg^{-1} = H$$

Then, H is said to be an *invariant sub-group* or *normal sub-group* or *normal divisor* of the group G. In other words, H is an invariant sub-group of G if all the conjugate sub-groups of H are equal to H.

As a first trivial example, the kernel ker(f) of a group homomorphism $f : G \to G'$ (i.e. the subset of elements of G having the neutral element E' of G' for image) is an invariant sub-group of G. Second, for any direct product $G \times G'$, the groups G and G' are invariant sub-groups of $G \times G'$.

The notion of normal divisor, introduced by Évariste Galois, leads to the important notions of *simple* and *semi-simple groups* (not considered in the present book) and to the notions of *factor group* or *quotient group* (see below).

PROPOSITION 5.2.– Let (H, τ) be a normal sub-group of a group (G, τ) and let us define the product of two cosets aH and bH of G by

$$aHbH = a\tau bH$$

Then, the set $\{H, aH, bH, \cdots\}$ of the different cosets of G with respect to H, endowed with the multiplication of cosets, forms a group called the *quotient group* or *factor group* of G by H, denoted as G/H.

Clearly, H is the neutral element of G/H and $a^{-1}H$ the inverse of aH. Note that $G \times G'/G \simeq G'$ and $G \times G'/G' \simeq G$.

5.2.11. *Abstract group - group table*

An abstract group is a group for which the nature of its elements and the group law are not explicitly given. A finite group (G, τ) can be given by its group table (called the multiplication table when $\tau = \times$ or the addition table when $\tau = +$): the set of elements $a\tau b$, for all a and b in G, can be displayed in a $|G|$ by $|G|$ array, the *group table* or *Cayley table* of G. The number of abstract finite groups of a given order, that are not isomorphic, is finite. As an illustration, Table 5.4 gives the number of abstract finite groups for some low order.

Order of the group	1	2	3	4	5	6	7	8	9	10	11	12	13
Number of groups	1	1	1	2	1	2	1	5	2	2	1	5	1

Table 5.4. *Number of (not isomorphic) abstract finite groups of low order*

C_2	e	a
e	e	a
a	a	e

Table 5.5. *Group table for the abstract group (G, τ) of order 2; the element at the intersection of the line x and the column y is $x\tau y$ (the table is symmetrical with respect to the diagonal of the table since the group is commutative); the elements of G are a and $e = a^2$ so that the group (G, τ) is cyclic (isomorphic to C_2)*

From Table 5.4, there is only one group table for $|G| = 2$ or 3 and two group tables for $|G| = 4$. This means that there is only one possibility for a group of order 2 or 3 (all groups of order 2 or 3 are isomorphic) and that there are two possibilities for a group of order 4. The corresponding group tables are given in Table 5.5 for $|G| = 2$, in Table 5.6 for $|G| = 3$ and in Tables 5.7 and 5.8 for $|G| = 4$. All groups of order 2 or 3 are Abelian and cyclic. There are two families of groups of order 4: in one family, all the groups are Abelian and cyclic, whereas in the other family, the groups are Abelian but not cyclic. The family of Abelian but not cyclic groups correspond to the abstract group referred to as the Klein four-group V (called *Vierergruppe* in German). It can be checked that V is isomorphic to the direct product $C_2 \times C_2$, where C_2 stands for the cyclic group of order 2.

C_3	e	a	b
e	e	a	b
a	a	b	e
b	b	e	a

Table 5.6. *Group table for the abstract group (G, τ) of order 3; the element at the intersection of the line x and the column y is $x\tau y$ (the table is symmetrical with respect to the diagonal of the table since the group is commutative); the elements of G are a, $b = a^2$ and $e = a^3$ so that the group (G, τ) is cyclic (isomorphic to C_3)*

C_4	e	a	b	c
e	e	a	b	c
a	a	b	c	e
b	b	c	e	a
c	c	e	a	b

Table 5.7. *Group table for the cyclic abstract group (G, τ) of order 4; the element at the intersection of the line x and the column y is $x\tau y$ (the table is symmetrical with respect to the diagonal of the table since the group is commutative); the elements of G are a, $b = a^2$, $c = a\tau b = a^3$ and $e = b^2 = a^4$ so that the group (G, τ) is cyclic (isomorphic to C_4)*

V	e	a	b	c
e	e	a	b	c
a	a	e	c	b
b	b	c	e	a
c	c	b	a	e

Table 5.8. *Group table for the not cyclic abstract group (G, τ) of order 4; the element at the intersection of the line x and the column y is $x\tau y$ (the table is symmetrical with respect to the diagonal of the table since the group is commutative); the elements of G are a, b, $c = a\tau b$ and $e = a^2$ so that the group (G, τ) is not cyclic; the abstract group corresponding to this table is called the Klein four-group generally denoted as V*

As realizations of the abstract group of order 3, let us mention the group of rotations leaving an equilateral triangle invariant (rotations around an axis perpendicular to the triangle, the group law being the composition of rotations) and the group $(\mathbb{Z}_3, +)$ of residues modulo 3 (the law group being the addition modulo 3). Along this vein, a realization of the Klein four-group V is the group of rotations leaving a rhombus invariant and realizations of the cyclic group C_4 are the group of rotations leaving a square invariant and the group $(\mathbb{Z}_4, +)$.

5.2.12. *Examples of groups*

Although this book is mainly concerned with finite groups, we start with three infinite (continuous) groups of relevance to Chapter 4.

5.2.12.1. *The special linear group SL$(n, \mathbb{C})$*

The set of $n \times n$ invertible matrices ($n = 2, 3, \cdots$) with coefficients in $\mathbb{C}$ endowed with the matrix multiplication constitutes a continuous group (in fact, a Lie group) referred to as the general linear group in n dimensions on

$\mathbb{C}$ and denoted as $\mathrm{GL}(n, \mathbb{C})$. The restriction of $\mathrm{GL}(n, \mathbb{C})$ to the matrices of determinant equal to 1 is called the special linear group in n dimensions on $\mathbb{C}$ and denoted as $\mathrm{SL}(n, \mathbb{C})$. Obviously, $\mathrm{SL}(n, \mathbb{C})$ is a sub-group of $\mathrm{GL}(n, \mathbb{C})$. In fact, the group $\mathrm{SL}(n, \mathbb{C})$ is an invariant sub-group of $\mathrm{GL}(n, \mathbb{C})$.

The limit case $\mathrm{GL}(1, \mathbb{C})$ corresponds to the group $(\mathbb{C}^*, \times)$, where the law $\times$ is the multiplication of complex numbers.

5.2.12.2. *The special unitary group SU*$(n, \mathbb{C})$

The set of $n \times n$ unitary matrices ($n = 2, 3, \cdots$) with coefficients in $\mathbb{C}$ endowed with the matrix multiplication constitutes a continuous group (in fact, a Lie group) referred to as the unitary group in n dimensions on $\mathbb{C}$ and denoted as $\mathrm{U}(n, \mathbb{C})$ or simply $\mathrm{U}(n)$. The restriction of $\mathrm{U}(n, \mathbb{C})$ to the matrices of determinant equal to 1 is called the special unitary group in n dimensions on $\mathbb{C}$ and denoted as $\mathrm{SU}(n, \mathbb{C})$ or simply $\mathrm{SU}(n)$. Obviously, $\mathrm{SU}(n, \mathbb{C})$ is a sub-group of $\mathrm{U}(n, \mathbb{C})$. We have the chains of groups

$$\mathrm{SU}(n, \mathbb{C}) \subset \mathrm{SL}(n, \mathbb{C}) \subset \mathrm{GL}(n, \mathbb{C}), \; \mathrm{SU}(n, \mathbb{C}) \subset \mathrm{U}(n, \mathbb{C}) \subset \mathrm{GL}(n, \mathbb{C})$$

The group $\mathrm{SU}(n, \mathbb{C})$ is an invariant sub-group of $\mathrm{U}(n, \mathbb{C})$.

5.2.12.3. *The special orthogonal group SO*$(n, \mathbb{R})$

The set of $n \times n$ orthogonal matrices ($n = 2, 3, \cdots$) with coefficients in $\mathbb{R}$ endowed with the matrix multiplication constitutes a continuous group (in fact, a Lie group) referred to as the orthogonal group in n dimensions on $\mathbb{R}$ and denoted as $\mathrm{O}(n, \mathbb{R})$ or simply $\mathrm{O}(n)$. The restriction of $\mathrm{O}(n, \mathbb{R})$ to the matrices of determinant equal to 1 is called the special orthogonal group in n dimensions on $\mathbb{R}$ and denoted as $\mathrm{SO}(n, \mathbb{R})$ or simply $\mathrm{SO}(n)$. Obviously, $\mathrm{SO}(n, \mathbb{R})$ is a sub-group of $\mathrm{O}(n, \mathbb{R})$. We have the chain of groups

$$\mathrm{SO}(n, \mathbb{R}) \subset \mathrm{O}(n, \mathbb{R}) \subset \mathrm{U}(n, \mathbb{C}) \subset \mathrm{GL}(n, \mathbb{C})$$

The group $\mathrm{SO}(n, \mathbb{R})$ is an invariant sub-group of $\mathrm{O}(n, \mathbb{R})$.

Note that SO(3) is isomorphic to the quotient group SU(2)/$\mathbb{Z}_2$, where $\mathbb{Z}_2$ stands here for the sub-group of SU(2) consisting of the two matrices

$$\begin{pmatrix} 1 & 0 \\ 0 & 1 \end{pmatrix}, \quad \begin{pmatrix} -1 & 0 \\ 0 & -1 \end{pmatrix}$$

(this sub-group is isomorphic to $(\mathbb{Z}_2, +)$). Furthermore, note that the groups SU(2) and SO(3) have the same Lie algebra.

In the three preceding examples, the fields $\mathbb{R}$ and $\mathbb{C}$ can be replaced by another field $\mathbb{K}$ yielding other groups, i.e. $\mathrm{GL}(n, \mathbb{K})$, $\mathrm{SL}(n, \mathbb{K})$, etc.

5.2.12.4. *The symmetric group S_n*

The set of permutations on n objects is a group with respect to the product (successive application) of permutations. This finite group, called the symmetric group on n objects and denoted as S_n, possesses $n!$ elements.

The group S_n comprises two types of elements: even permutations (products of an even number of transpositions) and odd permutations (products of an odd number of transpositions). The $\frac{1}{2}n!$ even permutations give rise to a sub-group of S_n, namely, the alternating group on n objects denoted as A_n. The group A_n is an invariant sub-group of S_n and the quotient group S_n/A_n is isomorphic to S_2.

One of the mathematical interests of S_n lies in the *Cayley theorem*, according to which each finite group of order n is isomorphic to a sub-group of the symmetric group S_n.

5.2.12.5. *The cyclic group C_d*

The abstract set $\{a, a^2, \cdots, a^d = e\}$, with $d \geq 1$, equipped with the law τ such that

$$a^n \tau a^m = a^{n+m \text{ mod d}}, \quad 1 \leq n \leq d, \quad 1 \leq m \leq d$$

is a cyclic group of order d and denoted as C_d.

A geometrical realization of the abstract group C_d is as follows. Let us consider a regular plan polygon with d sides. A rotation of angle $\frac{2\pi}{d}$ around the symmetry axis of the polygon leaves the polygon unchanged. The set of the d rotations of angles $k\frac{2\pi}{d}$, $1 \leq k \leq d$ or $0 \leq k \leq d-1$, leaving the polygon unchanged is a group with respect to the product (successive application) of rotations. This group is isomorphic to C_d.

The set of the d d-th roots of unity endowed with the product of complex numbers constitutes a second realization of the cyclic group C_d. This group is isomorphic to C_d.

A third realization of C_d is provided by the group $(\mathbb{Z}_d, +)$ corresponding to the set

$$\mathbb{Z}_d = \{0, 1, \cdots, d-1\}$$

endowed with the addition + modulo d (remember, $a + b$ modulo d is equal to the rest of the division of $a + b$ by d). This group is isomorphic to C_d.

Tables 5.5, 5.6 and 5.7 correspond to the group tables of the cyclic groups C_2, C_3 and C_4, respectively.

5.2.12.6. *The group of quaternions*

The set $\{1, -1, \mathrm{i}, -\mathrm{i}, \mathrm{j}, -\mathrm{j}, \mathrm{k}, -\mathrm{k}\}$ of Hamilton's quaternions endowed with the multiplication of quaternions (see 1.2.5.9) forms a group of order 8 called the quaternion group. All sub-groups of the quaternion group are invariant.

5.2.13. *Representations of a group*

5.2.13.1. *Linear representation of a group*

DEFINITION 5.10.– An n-dimensional complex linear representation D of a group (G, τ) is an homomorphic image of (G, τ) into $\mathrm{GL}(n, \mathbb{C})$. In other words, we have

$$D : (G, \tau) \to \mathrm{GL}(n, \mathbb{C})$$
$$a \mapsto D(a)$$

such that

$$\forall a \in G, \ \forall b \in G : D(a \tau b) = D(a) D(b)$$

where $D(a)D(b)$ is the matrix product of $D(a)$ by $D(b)$.

According to a tradition largely used in physics, we use D for Darstellung (which means representation in German).

The dimension of each of the matrices $D(a)$ is called the dimension of the representation D. The representation D is said to be unitary if each of the matrices $D(a)$ is unitary.

5.2.13.2. *Equivalent and non-equivalent representations*

DEFINITION 5.11.– Two representations D and D', of the same dimension n, of a group (G, τ) are said to be equivalent if there exists a $n \times n$ invertible matrix such that

$$\forall a \in G : D'(a) = M^{-1}D(a)M$$

where M^{-1} is the inverse of M. In the opposite case, the two representations D and D' are non-equivalent.

5.2.13.3. *Reducible and irreducible representations*

DEFINITION 5.12.– Let D be a linear representation of a finite (or compact continuous) group. If D possesses invariant sub-spaces, then D is said to be a *reducible representation*. The representation is *irreducible* in the opposite case.

From a practical point of view, a representation is reducible if it is equivalent to a representation where each matrix is a direct sum of sub-matrices, i.e. a block form along the diagonal, identical for each element of the group.

5.2.13.4. *Characters of a group*

DEFINITION 5.13.– Let D be a linear representation of a group (G, τ). The character $\chi^D(a)$, in the representation D, of the element a of G is defined by

$$\chi^D(a) = \mathrm{tr}\,(D(a))$$

where $\mathrm{tr}\,(D(a))$ is the trace of the matrix $D(a)$.

Note that the notation $\mathrm{tr}(M)$ for the trace of a matrix M should not be confused with $\mathrm{Tr}(x)$ for the trace of an element x of a ring or a field.

5.2.13.5. *Miscellaneous results*

– Each group G admits a trivial representation, called the *identity representation* of dimension 1 denoted as D^0, in which each element of the group is represented by 1 ($D^0(a) = \chi^{D^0}(a) = 1$, for all a in G).

– Each representation of a finite group is equivalent to a unitary representation. (This result is also valid for a compact continuous group.)

– In a given linear representation D of a group G, all the elements of a single conjugate class have the same character. This result can be described by

$$\chi^D(a) = \chi^D(\mathcal{C}_a)$$

where $\mathcal{C}_a$ denotes the conjugate class of the element a of G.

– A finite group possesses a finite number of non-equivalent irreducible representations. (A compact continuous group possesses a countable infinite number of non-equivalent irreducible representations.) The number of non-equivalent irreducible representations of a finite group is equal to the number of conjugate classes.

– For an Abelian finite group G, the number N of non-equivalent irreducible representations is $N = |G|$.

– The dimensions ℓ_i of the N non-equivalent irreducible representations of a finite group G satisfy the equality (*Burnside theorem*)

$$\sum_{i=1}^{N} \ell_i^2 = |G|$$

As an immediate corollary, the $N = |G|$ non-equivalent irreducible representations of an Abelian finite group G are all of dimension 1.

5.2.14. *Orthogonality relations*

5.2.14.1. *Orthogonality relations for finite groups*

PROPOSITION 5.3.– Let D^μ and D^ν be two irreducible unitary representations (either identical or non-equivalent) of a finite group G. The characters $\chi^{D^\mu}(\mathcal{C}_a)$, noted as $\chi^\mu(\mathcal{C}_a)$ for simplicity, satisfy the orthogonality relation

$$\sum_{\mathcal{C}_a} N_a \overline{\chi^\mu(\mathcal{C}_a)} \chi^\nu(\mathcal{C}_a) = |G| \, \delta[\mu, \nu] \qquad [5.1]$$

and its dual relation

$$\sum_{\mu} \overline{\chi^{\mu}(\mathcal{C}_a)}\chi^{\mu}(\mathcal{C}_b) = |G| \, \delta[\mathcal{C}_a, \mathcal{C}_b] \qquad [5.2]$$

where the bar indicates complex conjugation. In equation [5.1], N_a is the number of elements in the class $\mathcal{C}_a$ containing a, the summation on $\mathcal{C}_a$ runs over all the conjugate classes of G and $\delta[\mu, \nu] = 1$ or 0 according to which the representations D^{μ} and D^{ν} are identical or non-equivalent, respectively. In equation [5.2], the summation on μ runs over all the non-equivalent irreducible representations of G and $\delta[\mathcal{C}_a, \mathcal{C}_b] = 1$ or 0 according to which the conjugate classes $\mathcal{C}_a$ and $\mathcal{C}_b$ are identical or disjoint, respectively.

5.2.14.2. *Orthogonality relations for Abelian finite groups*

PROPOSITION 5.4.– In the particular case of an Abelian finite group G, all the irreducible representations are one-dimensional. Therefore, the orthogonality relations of G read

$$\sum_{a \in G} \overline{\chi^{\mu}(a)}\chi^{\nu}(a) = |G| \, \delta[\mu, \nu]$$

and

$$\sum_{\mu} \overline{\chi^{\mu}(a)}\chi^{\mu}(b) = |G| \, \delta(a, b)$$

which, like equations [5.1] and [5.2], reflect a duality between conjugate classes and irreducible representations of a finite group.

Note that, according to the tradition in theoretical physics, here we use the notation $\chi^{\mu}(a)$ for the character of the element a of G in the irreducible representation D^{μ} whereas, in field and ring theories, $\chi_y(x)$ and $\psi_k(x)$ stand for the characters of the element x in $\mathbb{GF}(p^m)$ or $\mathbb{GR}(p^s, m)$ in the additive and multiplicative character vectors χ_y and ψ_k, respectively.

5.2.14.3. *Character table of a finite group*

The various characters corresponding to all the non-equivalent irreducible representations of a finite group can be arranged into a table called the *character table* of the group. For an arbitrary finite group G, the character table is an N by N array, where N is the number of non-equivalent irreducible representations of G: the columns are labeled by the N conjugate classes of G and the lines by the N non-equivalent irreducible representations of G.

Note that two isomorphic groups have the same character table but two groups may have the same character table without being isomorphic.

Since this book is mainly concerned with finite rings and finite fields, we continue with some examples of character tables for Abelian finite groups. For an Abelian finite group (cyclic or not cyclic) G, the character table of G is a $|G|$ by $|G|$ array. The $|G|$ line-vectors of this $|G|$ by $|G|$ character table form a complete system of $|G|$ orthogonal vectors on the space $\mathbb{C}^{|G|}$, each line-vector being normalized to $\sqrt{|G|}$. A similar result holds for the $|G|$ column-vectors of the character table of G.

Tables 5.9, 5.10, 5.11 and 5.12 give the character tables for the groups V (the Klein four-group), C_2, C_3 and C_4 (three cyclic groups), respectively.

$x \in V \rightarrow$ $\chi^y \downarrow$	$e = a^2 = b^2$	a	b	$c = ab$
χ^0	1	1	1	1
χ^1	1	1	-1	-1
χ^2	1	-1	-1	1
χ^3	1	-1	1	-1

Table 5.9. *Character table of the Klein four-group V: the character at the intersection of the line χ^y and the column x is $\chi^y(x)$*

$x \in C_2 \rightarrow$ $\chi^y \downarrow$	$e = a^2$	a
χ^0	1	1
χ^1	1	-1

Table 5.10. *Character table of the cyclic group C_2: the character at the intersection of the line χ^y and the column x is $\chi^y(x)$*

$x \in C_3 \rightarrow$ $\chi^y \downarrow$	$e = a^3$	a	a^2
χ^0	1	1	1
χ^1	1	ω	ω^2
χ^2	1	ω^2	ω

Table 5.11. *Character table of the cyclic group C_3: the character at the intersection of the line χ^y and the column x is $\chi^y(x)$; note that $\omega = \mathrm{e}^{\mathrm{i}\frac{2\pi}{3}}$ and ω^2 are primitive roots of unity (of order 3)*

$x \in C_4 \rightarrow$ $\chi^y \downarrow$	$e = a^4$	a	$b = a^2$	$c = a^3$
χ^0	1	1	1	1
χ^1	1	-1	1	-1
χ^2	1	i	-1	$-\mathrm{i}$
χ^3	1	$-\mathrm{i}$	-1	i

Table 5.12. *Character table of the cyclic group C_4: the character at the intersection of the line χ^y and the column x is $\chi^y(x)$; note that i and $-\mathrm{i}$ are primitive roots of unity (of order 4)*

More generally, for the character table of an arbitrary cyclic finite group C_d of order d, it is appropriate to treat separately the case $d = 2n + 1$ odd with $C_{2n+1} = \{e, a, a^2, \cdots, a^{2n}\}$ and the case $d = 2n$ even with $C_{2n} = \{e, a, a^2, \cdots, a^{2n-1}\}$. In detail, we have

$$\chi^y(a^k) = \mathrm{e}^{\mathrm{i}\frac{2\pi}{2n+1}yk}, \quad k = 1, 2, \cdots, 2n+1$$

$$\chi^y = \chi^0, \chi^1, \chi^{-1}, \cdots, \chi^n, \chi^{-n}$$

for the cyclic group C_{2n+1}, and

$$\chi^y(a^k) = \mathrm{e}^{\mathrm{i}\frac{\pi}{n}yk}, \quad k = 1, 2, \cdots, 2n$$

$$\chi^y = \chi^0, \chi^1, \chi^{-1}, \cdots, \chi^n = \chi^{-n}$$

for the cyclic group C_{2n} (with $n \in \mathbb{N}_1$ in both cases). The character table of C_d is given in Table 5.13 for $d = 2n + 1$ and in Table 5.14 for $d = 2n$, respectively.

Tables 5.10 and 5.12 and Table 5.11 are the particular cases of Table 5.14 and Table 5.13, respectively (up to a rearrangement and a relabeling of some χ^y).

$x \in C_{2n+1} \rightarrow$ $\chi^y \downarrow$	$e = a^{2n+1}$	a	a^2	$\cdots$	a^{2n}
χ^0	1	1	1	$\cdots$	1
χ^1	1	ω	ω^2	$\cdots$	ω^{2n}
χ^{-1}	1	ω^{-1}	ω^{-2}	$\cdots$	ω^{-2n}
χ^2	1	ω^2	ω^4	$\cdots$	ω^{4n}
χ^{-2}	1	ω^{-2}	ω^{-4}	$\cdots$	ω^{-4n}
$\vdots$	$\vdots$	$\vdots$	$\vdots$	$\vdots$	$\vdots$
χ^n	1	ω^n	ω^{2n}	$\cdots$	ω^{2n^2}
χ^{-n}	1	ω^{-n}	ω^{-2n}	$\cdots$	ω^{-2n^2}

Table 5.13. *Character table of the cyclic group C_{2n+1} with the generator a: the character at the intersection of the line χ^y and the column x is $\chi^y(x)$; note that $\omega = e^{i\frac{2\pi}{2n+1}}$ is a primitive root of unity (of order $2n+1$)*

$x \in C_{2n} \rightarrow$ $\chi^y \downarrow$	$e = a^{2n}$	a	a^2	$\cdots$	a^{2n-1}
χ^0	1	1	1	$\cdots$	1
χ^1	1	ω	ω^2	$\cdots$	ω^{2n-1}
χ^{-1}	1	ω^{-1}	ω^{-2}	$\cdots$	ω^{-2n+1}
χ^2	1	ω^2	ω^4	$\cdots$	$\omega^{2(2n-1)}$
χ^{-2}	1	ω^{-2}	ω^{-4}	$\cdots$	$\omega^{-2(2n-1)}$
$\vdots$	$\vdots$	$\vdots$	$\vdots$	$\vdots$	$\vdots$
$\chi^n = \chi^{-n}$	1	ω^n	ω^{2n}	$\cdots$	$\omega^{n(2n-1)}$

Table 5.14. *Character table of the cyclic group C_{2n} with the generator a: the character at the intersection of the line χ^y and the column x is $\chi^y(x)$; note that $\omega = e^{i\frac{\pi}{n}}$ is a primitive root of unity (of order $2n$)*

Bibliography

Mathematical literature: rings and fields

[ART 98] ARTIN E., *Galois Theory*, Dover, New York, 1998.

[CAR 81] CARREGA J.C., *Théorie des corps; la règle et le compas*, Hermann, Paris, 1981.

[DEM 08] DEMAZURE M., *Cours d'algèbre*, Cassini, Paris, 2008.

[GOZ 97] GOZARD I., *Théorie de Galois*, Ellipses, Paris, 1997.

[LID 08] LIDL R., NIEDERREITER H., *Finite Fields*, Cambridge University Press, Cambridge, 2008.

[MCD 74] MCDONALD B.R., *Finite Rings with Identity*, M. Dekker, New York, 1974.

[MCE 87] MCELIECE R.J., *Finite Fields for Computer Scientists and Engineers*, Kluwer, Boston, 1987.

[STE 95] STEWART I., *Galois Theory*, Chapman & Hall, London, 1995.

[VAN 53] VAN DER WAERDEN B.L., *Modern Algebra*, Volume 1, Frederick Ungar, New York, 1953.

[WAN 03] WAN Z.-X., *Lectures on Finite Fields and Galois Rings*, World Scientific, Hackensack, 2003.

Mathematical literature: group theory

[BAL 86] BALIAN R., ITZYKSON C., "Observations sur la mécanique quantique finie", *Comptes Rendus des Séances de l'Académie des Sciences*, vol. 303, pp. 773–778, 1986. [Weyl pair and Heisenberg-Weyl group]

[HAM 89] HAMERMESH M., *Group Theory and its Application to Physical Problems*, Dover, New York, 1989. [Group theory for physicists]

[KOS 94] KOSTRIKIN A.I., TIEP P.H., *Orthogonal Decompositions and Integral Lattices*, Walter de Gruyter, Berlin, 1994. [Orthogonal decompositions of complex simple Lie algebras]

[PAT 88] PATERA J., ZASSENHAUS H., "The Pauli matrices in n dimensions and finest gradings of simple Lie algebras of type A_{n-1}", *Journal of Mathematical Physics*, vol. 29, pp. 665–673, 1988. [Generalized Pauli matrices, Pauli group, orthogonal decomposition of $\mathrm{sl}(d, \mathbb{C})$]

[ŠŤO 84] ŠŤOVÍČEK P., TOLAR J., "Quantum mechanics in a discrete space–time", *Reports on Mathematical Physics*, vol. 20, pp. 157–170, 1984. [Weyl pair and Heisenberg-Weyl group]

[WIG 59] WIGNER E.P., *Group Theory and its Application to the Quantum Mechanics of Atomic Spectra*, Academic Press, New York, 1959. [Group theory for physicists]

[WYB 74] WYBOURNE B.G., *Classical Groups for Physicists*, Wiley, New York, 1974. [Group theory for physicists]

Mathematical literature: number theory

[BER 81] BERNDT B.C., EVANS R.J., "The determination of Gauss sums", *Bulletin of the American Mathematical Society*, vol. 5, pp. 107–129, 1981.

[BER 98] BERNDT B.C., EVANS R.J., WILLIAMS K.S., *Gauss and Jacobi Sums*, Wiley, New York, 1998.

[HUA 82] HUA L.K., *Introduction to Number Theory*, Springer-Verlag, Berlin, 1982.

[LAN 70] LANG S., *Algebraic Number Theory*, Addison-Wesley, Reading, 1970.

[WEI 48] WEIL A., "On some exponential sums", *Proceedings of the National Academy of Sciences of the United States of America*, vol. 34, pp. 204–207, 1948.

Theoretical physics literature: MUBs

[ALB 09a] ALBOUY O., Discrete algebra and geometry applied to the Pauli group and mutually unbiased bases in quantum information theory, PhD thesis, University of Lyon, 2009. [Pauli group, construction of MUBs via finite algebraic and geometrical methods]

[ALB 09b] ALBOUY O., "The isotropic lines of $\mathbb{Z}_d^2$", *Journal of Physics A: Mathematical and Theoretical*, vol. 42, pp. 072001–072009, 2009. [Discrete Wigner distributions and Lagrangian submodules]

[ALL 80] ALLTOP W.O., "Complex sequences with low periodic correlations", *IEEE Transactions on Information Theory*, vol. IT-26, pp. 350–354, 1980. [Construction of MUBs for $p \geq 5$, p prime]

[APP 09] APPLEBY D.M., "SIC-POVMS and MUBS: geometrical relationships in prime dimension", *AIP Conference Proceedings, Foundations of Probability and Physics-5*, Växjö, Sweden, vol. 1101, pp. 223–232, 2009. [Connection between MUBs and SIC-POVMs]

[ARA 03] ARAVIND P.K., "Solution to the King's problem in prime power dimensions", *Zeitschrift für Naturforschung*, vol. 58a, pp. 85–92, 2003. [MUBs and the Mean King problem]

[ARC 05] ARCHER C., "There is no generalization of known formulas for mutually unbiased bases", *Journal of Mathematical Physics*, vol. 46, p. 022106, 2005. [Formulas for MUBs via Galois rings]

[ASC 07] ASCHBACHER M., CHILDS A.M., WOCJAN P., "The limitations of nice mutually unbiased bases", *Journal of Algebraic Combinatorics*, vol. 25, pp. 111–123, 2007. [MUBs in composite dimension, MUBs and Hadamard matrices]

[ATA 10] ATAKISHIYEV N.M., KIBLER M.R., WOLF K.B., "SU(2) and SU(1,1) approaches to phase operators and temporally stable phase states: Applications to mutually unbiased bases and discrete Fourier transforms", *Symmetry*, vol. 2, pp. 1461–1484, 2010. [MUBs and temporally stable phase states]

[BAN 02] BANDYOPADHYAY S., BOYKIN P.O., ROYCHOWDHURY V. *et al.*, "A new proof for the existence of mutually unbiased bases", *Algorithmica*, vol. 34, pp. 512–528, 2002. [Existence and construction of MUBs based on classes of commuting operators and generalized Pauli matrices]

[BEC 00] BECHMANN-PASQUINUCCI H., PERES A., "Quantum cryptography with 3-state systems", *Physical Review Letters*, vol. 85, pp. 3313–3316, 2000. [MUBs and quantum cryptography]

[BEN 05a] BENGTSSON I., "MUBs, polytopes, and finite geometries", *AIP Conference Proceedings, Foundations of Probability and Physics-3*, Växjö, Sweden, vol. 750, pp. 63–69, 2005. [MUBs and finite affine planes]

[BEN 05b] BENGTSSON I., ERICSSON Å., "Mutually unbiased bases and the complementary polytope", *Open Systems and Information Dynamics*, vol. 12, pp. 107–120, 2005. [MUBs and polytope]

[BEN 07a] BENGTSSON I., "Three ways to look at mutually unbiased bases", *AIP Conference Proceedings, Foundations of Probability and Physics-4*, Växjö, Sweden, vol. 889, pp. 40–51, 2007. [Geometrical approach to MUBs]

[BEN 07b] BENGTSSON I., BRUZDA W., ERICSSON Å. *et al.*, "Mutually unbiased bases and Hadamard matrices of order six", *Journal of Mathematical Physics*, vol. 48, pp. 052106–052127, 2007. [MUBs in dimension 6, MUBs and Hadamard matrices]

[BJÖ 07] BJÖRK G., ROMERO J.L., KLIMOV A.B. *et al.*, "Mutually unbiased bases and discrete Wigner functions", *Journal of the Optical Society of America*, vol. 24, pp. 371–378, 2007. [MUBs for three-qubit systems]

[BOY 07] BOYKIN P.O., SITHARAM M., TIEP P.H. *et al.*, "Mutually unbiased bases and orthogonal decompositions of Lie algebras", *Quantum Information and Computation*, vol. 7, pp. 371–382, 2007. [MUBs and orthogonal decomposition of $\mathrm{sl}(d, \mathbb{C})$]

[BRI 08] BRIERLEY S., WEIGERT S., "Maximal sets of mutually unbiased quantum states in dimension six", *Physical Review A*, vol. 78, pp. 042312–042320, 2008. [MUBs in dimension 6]

[BRI 09] BRIERLEY S., WEIGERT S., "Constructing mutually unbiased bases in dimension six", *Physical Review A*, vol. 79, pp. 052316–052329, 2009. [MUBs in dimension 6, MUBs and Hadamard matrices]

[BUT 07] BUTTERLEY P., HALL W., "Numerical evidence for the maximum number of mutually unbiased bases in dimension six", *Physics Letters A*, vol. 369, pp. 5–8, 2007. [MUBs in dimension 6]

[CAL 97] CALDERBANK A.R., CAMERON P.J., KANTOR W.M. *et al.*, "$\mathbb{Z}_4$–Kerdock codes, orthogonal spreads, and extremal Euclidean line-sets", *Proceedings of the London Mathematical Society*, vol. 75, pp. 436–480, 1997. [Galois fields and Galois rings for codes in classical information theory]

[CER 02] CERF N.J., BOURENNANE M., KARLSSON A. *et al.*, "Security of quantum key distribution using d-level systems", *Physical Review Letters*, vol. 88, pp. 127902–127906, 2002. [MUBs and quantum cryptography, quantum cryptosystems based on qudits]

[CHA 02] CHATURVEDI S., "Aspects of mutually unbiased bases in odd prime power dimensions", *Physical Review A*, vol. 65, pp. 044301–044305, 2002. [MUBs and characters of cyclic groups]

[COM 09] COMBESCURE M., "Block-circulant matrices with circulant blocks, Weil sums, and mutually unbiased bases. II. The prime power case", *Journal of Mathematical Physics*, 50, pp. 032104–032116, 2009. [MUBs and circulant matrices]

[DAO 11] DAOUD M., KIBLER M.R., "Phase operators, phase states and vector phase states for SU_3 and $SU_{2,1}$", *Journal of Mathematical Physics*, vol. 52, pp. 082101–082122, 2011. [MUBs and phase states]

[DIŢ 10] DIŢĂ P., "Hadamard matrices from mutually unbiased bases", *Journal of Mathematical Physics*, vol. 51, pp. 072202–072222, 2010. [Construction of generalized Hadamard matrices from MUBs]

[DUR 04] DURT T., "If $1 = 2 + 3$, then $1 = 2.3$: Bell states, finite groups, and mutually unbiased bases, a unifying approach", *arXIV*, arXiv:0401046v2 [quant-ph], 2004. [Construction of MUBs]

[DUR 05] DURT T., "About mutually unbiased bases in even and odd prime power dimensions", *Journal of Physics A: Mathematical and General*, vol. 38, pp. 5267–5283, 2005. [Construction of MUBs via Galois fields]

[DUR 06] DURT T., "About the Mean King's problem and discrete Wigner distributions", *International Journal of Modern Physics B*, vol. 20, pp. 1742–1760, 2006. [MUBs and the Mean King problem]

[DUR 10] DURT T., ENGLERT B.-G., BENGTSSON I. *et al.*, "On mutually unbiased bases", *International Journal of Quantum Information*, vol. 8, pp. 535–640, 2010. [Review paper on MUBs]

[ENG 01] ENGLERT B.-G., AHARONOV Y., "The mean king's problem: prime degrees of freedom", *Physics Letters A*, vol. 284, pp. 1–5, 2001. [MUBs and the Mean King problem]

[GHI 13] GHIU I., "Generation of all sets of mutually unbiased bases for three-qubit systems", *Physica Scripta*, vol. T153, pp. 014027–014032, 2013. [MUBs for three-qubit systems and Galois fields]

[GIB 04] GIBBONS K.S., HOFFMAN M.J., WOOTTERS W.K., "Discrete phase space based on finite fields", *Physical Review A*, vol. 70, pp. 062101–062124, 2004. [Construction of MUBs via Galois fields, MUBs and finite-dimensional Wigner function]

[GOD 09] GODSIL C., ROY A., "Equiangular lines, mutually unbiased bases, and spin models", *European Journal of Combinatorics*, vol. 30, pp. 246–262, 2009. [MUBs and equiangular lines, spin models and Galois fields]

[GOY 13] GOYENECHE D., "Mutually unbiased triplets from non-affine families of complex Hadamard matrices in dimension 6", *Journal of Physics A: Mathematical and Theoretical*, vol. 46, pp. 105301–105316, 2013. [MUBs in dimension 6]

[GRA 05] GRASSL M., "Tomography of quantum states in small dimensions", *Electronic Notes in Discrete Mathematics*, vol. 20, pp. 151–164, 2005. [MUBs and quantum state tomography]

[GRA 09] GRASSL M., "On SIC-POVMs and MUBs in dimension 6", *Proceedings ERATO Conference on Quantum Information Science (EQIS'04)*, Tokyo, pp. 60–61, 2005 and arXiv:0406175v2 [quant-ph], 2009. [MUBs in dimension 6]

[HAY 05] HAYASHI A., HORIBE M., HASHIMOTO T., "Mean king's problem with mutually unbiased bases and orthogonal Latin squares", *Physical Review A*, vol. 71, pp. 052331–052335, 2005. [MUBs and the Mean King problem]

[HEA 11] HEALY J.J., WOLF K.B., "Discrete canonical transforms that are Hadamard matrices", *Journal of Physics A: Mathematical and Theoretical*, vol. 44, pp. 265302–265312, 2011. [MUBs and Hadamard matrices]

[HEA 06] HEATH R.W., STROHMER T., PAULRAJ A.J., "On quasi-orthogonal signatures for CDMA systems", *IEEE Transactions on Information Theory*, vol. 52, pp. 1217–1226, 2006. [MUBs and classical information theory]

[IVA 81] IVANOVIĆ I.D., "Geometrical description of quantal state determination", *Journal of Physics A: Mathematical and General*, vol. 14, pp. 3241–3245, 1981. [Maximum number of MUBs, construction of MUBs in odd prime dimension]

[KAL 13] KALEV A., "A geometrical relation between symmetric operators and mutually unbiased operators", *arXIV*, arXiv:1305.6044 [quant-ph], 2013. [MUBs and finite plane geometry]

[KIB 06a] KIBLER M.R., "Angular momentum and mutually unbiased bases", *International Journal of Modern Physics B*, vol. 20, pp. 1792–1801, 2006. [MUBs and quantum theory of angular momentum]

[KIB 06b] KIBLER M.R., PLANAT M., "A SU(2) recipe for mutually unbiased bases", *International Journal of Modern Physics B*, vol. 20, pp. 1802–1807, 2006. [Construction of MUBs via SU(2)]

[KIB 08] KIBLER M.R., "Variations on a theme of Heisenberg, Pauli and Weyl", *Journal of Physics A: Mathematical and Theoretical*, vol. 41, pp. 375302–375321, 2008. [Weyl pairs, Heisenberg-Weyl group, Pauli group and unitary group]

[KIB 09] KIBLER M.R., "An angular momentum approach to quadratic Fourier transform, Hadamard matrices, Gauss sums, mutually unbiased bases, the unitary group and the Pauli group", *Journal of Physics A: Mathematical and Theoretical*, vol. 42, pp. 353001–353029, 2009. [Topical review about MUBs]

[KIB 14] KIBLER M.R., "On two ways to look for mutually unbiased bases", *Acta Polytechnica 54*, Prague, pp. 124–126, 2014. [MUBs and equiangular lines]

[KLA 04] KLAPPENECKER A., RÖTTELER M., "Constructions of mutually unbiased bases", *Lecture Notes in Computer Science*, vol. 2948, pp. 137–144, 2004. [Construction of MUBs via Galois fields and Galois rings]

[KLA 05] KLAPPENECKER A., RÖTTELER M., "Mutually unbiased bases are complex projective 2-designs", *Proceedings of the 2005 IEEE International Symposium on Information Theory*, Adelaide, Australia, pp. 1740–1744, 2005. [MUBs and projective 2-designs]

[KLI 05a] KLIMOV A.B., SÁNCHEZ-SOTO L.L., DE GUISE H., "Multicomplementary operators via finite Fourier transform", *Journal of Physics A: Mathematical and General*, vol. 38, pp. 2747–2760, 2005. [Construction of MUBs via Galois fields, two- and three-qubit systems, two-qutrit systems]

[KLI 05b] KLIMOV A.B., SÁNCHEZ-SOTO L.L., DE GUISE H., "A complementarity-based approach to phase in finite-dimensional quantum systems", *Journal of Optics B: Quantum and Semiclassical Optics*, vol. 7, pp. 283–287, 2005. [Construction of MUBs based on classes of commuting operators]

[KLI 07] KLIMOV A.B., ROMERO J.L., BJÖRK G. *et al.*, "Geometrical approach to mutually unbiased bases", *Journal of Physics A: Mathematical and Theoretical*, vol. 40, pp. 3987–3998, 2007 and vol. 40, p. 9177, 2007. [MUBs for two-qubit systems]

[KLI 08] KLIMOV A.B., MUÑOZ C., FERNÁNDEZ A. *et al.*, "Optimal quantum-state reconstruction for cold trapped ions", *Physical Review A*, vol. 77, p. 060303(R), 2008. [MUBs and quantum tomography]

[KON 01] KONIORCZYK M., BUŽEK V., JANSZKY J., "Wigner-function description of quantum teleportation in arbitrary dimensions and a continuous limit", *Physical Review A*, vol. 64, pp. 034301–034305, 2001. [MUBs and quantum teleportation]

[LAW 02] LAWRENCE J., BRUKNER Č., ZEILINGER A., "Mutually unbiased binary observable sets on N qubits", *Physical Review A*, vol. 65, pp. 032320–032325, 2002. [MUBs and generalized Pauli matrices for N-qubit systems]

[LAW 04] LAWRENCE J., "Mutually unbiased bases and trinary operator sets for N qutrits", *Physical Review A*, vol. 70, pp. 012302–012303, 2004. [MUBs for N qutrits and Pauli group]

[LAW 11] LAWRENCE J., "Entanglement patterns in mutually unbiased basis sets", *Physical Review A*, vol. 84, pp. 022338–022339, 2011. [MUBs and entanglement]

[MCN 12a] MCNULTY D., WEIGERT S., "The limited role of mutually unbiased product bases in dimension six", *Journal of Physics A: Mathematical and Theoretical*, vol. 45, pp. 102001–102006, 2012. [MUBs in dimension 6]

[MCN 12b] MCNULTY D., WEIGERT S., "All mutually unbiased product bases in dimension six", *Journal of Physics A: Mathematical and Theoretical*, vol. 45, pp. 135307–135329, 2012. [MUBs in dimension 6]

[OLU 16] OLUPITAN T., LEI C., VOURDAS A., "An analytic function approach to weak mutually unbiased bases", *Annals of Physics*, vol. 371, pp. 1–19, 2016. [Weak MUBs]

[PAZ 05] PAZ J.P., RONCAGLIA A.J., SARACENO M., "Qubits in phase space: Wigner-function approach to quantum-error correction and the mean-king problem", *Physical Review A*, vol. 72, pp. 012309–012328, 2005. [MUBs, quantum error correction and the Mean King problem]

[PIT 04] PITTENGER A.O., RUBIN M.H., "Mutually unbiased bases, generalized spin matrices and separability", *Linear Algebra and its Applications*, vol. 390, pp. 255–278, 2004. [Construction of MUBs via generalized Pauli matrices and Galois fields, separability of bases]

[PIT 05] PITTENGER A.O., RUBIN M.H., "Wigner functions and separability for finite systems", *Journal of Physics A: Mathematical and General*, vol. 38, pp. 6005–6036, 2005. [MUBs and discrete Wigner functions]

[PLA 06] PLANAT M., SANIGA M., KIBLER M.R., "Quantum entanglement and projective ring geometry", *Symmetry, Integrability and Geometry: Methods and Applications*, vol. 2, pp. 066–080, 2006. [Entanglement and Galois rings]

[PLA 10] PLANAT M., KIBLER M.R., "Unitary reflection groups for quantum fault tolerance", *Journal of Computational and Theoretical Nanoscience*, vol. 7, pp. 1759–1770, 2010. [Quantum computing and unitary reflections]

[ROM 05] ROMERO J.L., BJÖRK G., KLIMOV A.B. *et al.*, "Structure of the sets of mutually unbiased bases for N qubits", *Physical Review A*, vol. 72, pp. 062310–062318, 2005. [MUBs for N qubits]

[SÁN 06] SÁNCHEZ-SOTO L.L., KLIMOV A.B., DE GUISE H., "Multipartite quantum systems: phases do matter after all", *International Journal of Modern Physics B*, vol. 20, pp. 1877–1884, 2006. [MUBs, quantum phase and complementarity]

[SAN 04] SANIGA M., PLANAT M., ROSU H., "Mutually unbiased bases and finite projective planes", *Journal of Optics B: Quantum and Semiclassical Optics*, vol. 6, pp. L19–L20, 2004. [SPR conjecture for MUBs and finite projective planes]

[SAN 07] SANIGA M., PLANAT M., KIBLER M.R. *et al.*, "A classification of the projective lines over small rings", *Chaos, Solitons and Fractals*, vol. 33, pp. 1095–1102, 2007. [Projective lines over rings]

[SCH 60] SCHWINGER J., "Unitary operator bases", *Proceedings of the National Academy of Sciences of the United of America*, vol. 46, pp. 570–579, 1960. [Unitary operator bases as ancestors of MUBs]

[SHA 12] SHALABY M., VOURDAS A., "Weak mutually unbiased bases", *Journal of Physics A: Mathematical and Theoretical*, vol. 45, pp. 052001–052016, 2012. [Notion of weak MUBs]

[SPE 13] SPENGLER C., KRAUS B., "Graph-state formalism for mutually unbiased bases", *Physical Review A*, vol. 88, pp. 052323–052344, 2013. [Construction of MUBs via graph theory]

[ŠUL 07] ŠULC P., TOLAR J., "Group theoretical construction of mutually unbiased bases in Hilbert spaces of prime dimensions", *Journal of Physics A: Mathematical and Theoretical*, vol. 40, pp. 15099–15111, 2007. [MUBs in prime dimension via group theory]

[SVE 08] SVETLICHNY G., "Feynman's integral is about mutually unbiased bases", *arXIV*, arXiv:0708.3079v3 [quant-ph], 2008. [MUBs and the Feyman path integral formalism]

[TOL 09] TOLAR J., CHADZITASKOS G., "Feynman's path integral and mutually unbiased bases", *Journal of Physics A: Mathematical and Theoretical*, vol. 42, pp. 245306–245317, 2009. [MUBs and the Feyman path integral formalism]

[VOU 96] VOURDAS A., "The angle-angular momentum quantum phase space", *Journal of Physics A: Mathematical and General*, vol. 29, pp. 4275–4288, 1996. [Heisenberg-Weyl group and Galois field]

[VOU 05] VOURDAS A., "Galois quantum systems", *Journal of Physics A: Mathematical and General*, vol. 38, pp. 8453–8471, 2005. [finite quantum system and Galois field]

[VOU 06] VOURDAS A., "Galois quantum systems, irreducible polynomials and Riemann surfaces", *Journal of Mathematical Physics*, vol. 47, pp. 092104–092119, 2006. [Galois quantum systems]

[VOU 07] VOURDAS A., "Quantum systems in finite Hilbert space: Galois fields in quantum mechanics", *Journal of Physics A: Mathematical and Theoretical*, vol. 40, pp. R285–R331, 2007. [Galois fields in quantum mechanics]

[WOC 05] WOCJAN P., BETH T., "New construction of mutually unbiased bases in square dimensions", *Quantum Information and Computation*, vol. 5, pp. 93–101, 2005. [MUBs in composite dimension, Latin squares and Hadamard matrices]

[WOO 86] WOOTTERS W.K., "Quantum mechanics without probability amplitudes", *Foundations of Physics*, vol. 16, pp. 391–405, 1986. [Definition of MUBs, MUBs for $d = p$ odd prime]

[WOO 87] WOOTTERS W.K., "A Wigner function formulation of finite-state quantum mechanics", *Annals of Physics*, New York, vol. 176, pp. 1–21, 1987. [Definition of MUBs, MUBs for $d = p$ odd prime and finite-dimensional Wigner function]

[WOO 89] WOOTTERS W.K., FIELDS B.D., "Optimal state-determination by mutually unbiased measurements", *Annals of Physics*, New York, vol. 191, pp. 363–381, 1989. [Maximum number of MUBs, construction of MUBs via Galois fields]

[ZAU 99] ZAUNER G., Quantendesigns: Grundzüge einer nichtcommutativen Designtheorie, PhD Thesis, University of Vienna, 1999. [MUBs in dimension 6, conjecture $N(6) = 3$]

Useful web links

[ALO XX] ALOI N., Évariste Galois: film in French, available at: https://www.youtube.com/watch?v=JGEYOVhDwoc

[ASH XX] ASH R.B., Abstract algebra: the basic graduate year and a course in algebraic number theory (two online books), available at: http://www.math.uiuc.edu/~r-ash/

[AST XX] ASTRUC A., Évariste Galois, available at: https://www.youtube.com/watch?v=Sl2FBpkTGCc

[BRO XX] BROWNING T., Exponential sums over finite fields, available at: https://people.maths.bris.ac.uk/~matdb/tcc/EXP/

[BRU XX] BRUZDA W., TADEJ W., ŻYCZKOWSKI K., Online catalogue of known Hadamard matrices, available at: http://chaos.if.uj.edu.pl/~karol/hadamard/

[CHA XX] CHABAUD F., Polynomials over Galois fields, available at: http://fchabaud.free.fr/English/default.php?COUNT=1&FILE0=Poly

[CON XX] CONNES A., *La pensée d'Évariste Galois et le formalisme moderne*, available at: http://www.alainconnes.org/docs/galoistext.pdf

[DEM XX] DEMAZURE M., *Cours d'algèbre, compléments*, available at: http://www.cassini.fr/COMPAGNONS/Demazure/Demazure-complements.pdf

[LAS XX] LASLO Y., *Introduction à la théorie de Galois*, available at: http://www.cmls.polytechnique.fr/perso/laszlo/galois/galois.pdf

[MIL XX] MILNE J.S., Fields and Galois theory, available at: http://www.jmilne.org/math/CourseNotes/ft.html

[MUR XX] MURPHY T., Finite fields, available at: http://www.maths.tcd.ie/pub/Maths/Courseware/373-2000/FiniteFields.pdf

[RUS XX] RUSKEY F., Information on primitive and irreducible polynomials, available at: http://theory.cs.uvic.ca/inf/neck/PolyInfo.html

[SOL XX] SOLÉ P., Galois ring, Encyclopedia of Mathematics, available at: http://www.encyclopediaofmath.org/index.php?title=Galois_ring&oldid=14749

Index

E, F, G

H, I, K

R, S, T

U, V, W, Z

Printed in the United States
By Bookmasters